Saltmarsh

BRITISH WILDLIFE COLLECTION

5

Saltmarsh

Clive Chatters

BLOOMSBURY
LONDON · OXFORD · NEW YORK · NEW DELHI · SYDNEY

Dedication

To Dr Francis Rose (1921–2006), who got us thinking.

Half title: Thrift, Shetland.
Frontispiece: Common Sea-lavender on Warham Marsh, Norfolk.

Bloomsbury Natural History
An imprint of Bloomsbury Publishing Plc

50 Bedford Square	1385 Broadway
London	New York
WC1B 3DP	NY 10018
UK	USA

www.bloomsbury.com

BLOOMSBURY and the Diana logo are trademarks of Bloomsbury Publishing Plc

First published 2017

British Library Cataloguing-in-Publication Data
A catalogue record for this book is available from the British Library.

ISBN HB: 978-1-4729-3359-1
ePDF: 978-1-4729-4297-5
ePub: 978-1-4729-3360-7

2 4 6 8 10 9 7 5 3 1

Page layouts by Susan McIntyre
Jacket artwork by Carry Akroyd

MIX
Paper from
responsible sources
FSC® C104723

To find out more about our authors and books visit www.bloomsbury.com. Here you will find extracts,
author interviews, details of forthcoming events and the option to sign up for our newsletters.

Contents

Preface   7

1 Introduction   12

2 Far from the shore   34

3 On the wind   58

4 In the Highlands   66

5 Atlantic gateway   88

6 The merse of the Solway Firth   106

7 Bae Ceredigion   124

8 Genesis of the Humber   136

9 Seawalls and the Severn   150

10 Capital marsh country   172

11 The legacy of the Solent's saltworks   190

12 Southampton's Spartinas   200

13 Conservation before conservationists   220

14 The advent of nature conservation   232

15 Modest proposals   242

16 The rise of regulation   260

17 International perspectives   286

18 Invasive and non-native species   296

19 Conservation in practice   308

20 Rejuvenation   328

21 Time and tide   342

Appendix A: A provisional inventory of
inland saltmarshes in Britain   346

Appendix B: Special Areas of Conservation in
Britain supporting saltmarsh habitats   355

References and further reading   361

Species names   368

Illustration credits   372

Index   374

Preface

Saltmarshes have formed the backdrop of much of my life. I was born in a Tudor cottage by the church of Saint Leonard-at-the-Hythe. The ancient settlement stood comfortably above the tidal Colne where the Romans' quay served their town of Camulodunum, modern-day Colchester in Essex. I have been told, but have no recollection, of being aired in a pram along those seawalls. Both of my parents had family associations with the saltmarsh country of the Colne and Blackwater. Childhood was interspersed with visits to Mersea Island synchronised with the tides lest the Strood, the only link to the mainland, became impassable. Trips to Mersea were accompanied by stories of the war years on a front-line island, the great earthquake and the terrible floods of '53; a chronology muddled in my childish mind. Amongst the stories were romantic concoctions of marshes populated by ghostly Roman legions and of Black Shuck, the devil dog, hunting down souls to their doom. My father encouraged us to sort fact from fable and to celebrate what gave the Island its true character. The Essex earthquake of 1884 was evidenced by a fragment of mirror; the original thrown from the mantelpiece by the tremors, with the remnant remounted and backed with press cuttings. Apples were stored in the bomb shelter my grandfather had dug in the garden and everyone over a certain age had a story about '53. My enduring memory of Mersea is a pervasive oily reek; even the beach where we played was a sandy veneer over anoxic mud. The tang of the marsh travelled home in our skin and even today releases remembrances of things past.

I had the great fortune to attend Wye College, the school of London University set in the Downs of East Kent. It was here I was lectured in the discipline of ecology by Bryn Green, a practical conservationist before he took to academia. Bryn introduced us to the dissatisfaction amongst some ecologists in the current thinking on the origins of Britain's wildlife. The dominant model at the time was that promoted by the Cambridge Botany School as exemplified

OPPOSITE PAGE:
St Thomas's Church,
Fairfield, Kent.

by the work of Sir Harry Godwin. Bryn encouraged us to question whether the diverse wildlife that surrounded the college really developed in a handful of millennia following the clearing of dense forests by farmers. Francis Rose, the Reader in Biogeography at King's College, London, joined us as a guest lecturer. Francis had an unrivalled knowledge of the vegetation of north-west Europe on which he drew to challenge our thinking; he developed and dismissed hypotheses relating to theoretical refugia where the species of open habitats survived the boreal blanket. His realisation of the integral role of large wild herbivores anticipated the conclusions of continental ecologists such as Frans Vera by several decades.

Wye prided itself in mixing the theoretical with the practical and was ideally located for a budding naturalist. The college estate included one of the richest orchid sites in Britain; within walking distance there were ancient woods, a peat bog and the wetlands of the Stour. At this time the conservation of the wildlife we enjoyed was supposedly to be achieved through the voluntary principle. That principle was clearly not working and during my short years in Kent I witnessed both the dereliction and destruction of superb habitats including parts of the nearby Romney Marshes being drained and ploughed. All was not lost; there were still sanctuaries to enjoy and learn from. In one such site, not far from Brookland, Brackish Water-crowfoot *Ranunculus baudotii* filled the fleets surrounding St Thomas's Church. I recall my puzzlement at the context; what was Brackish Water-crowfoot doing in a place so far from the sea? I had much to find out about the history of our landscape.

After college, chance took me to the Isle of Wight where volunteering for the National Trust gave me simple accommodation at Newtown. The Newtown River is the most complete of the Solent estuaries in that there are no major settlements, few incursions on the intertidal and most of the land is gently farmed. The saltmarshes at Newtown formed part of an extensive landscape of exceptional quality with transitional habitats supporting many exacting species. There was Marsh-mallow *Althaea officinalis* in the shady fringes of the estuarine woods, Slender Hare's-ear *Bupleurum tenuissimum* on the stump of a neglected seawall, and old records of Foxtail Stonewort *Lamprothamnium papulosum* submerged in the lagoons of an abandoned saltworks. Curlew feeding on the intertidal would roost on the ancient meadows, butterflies breeding on those meadows happily nectared on the marsh, the wildlife did not classify the habitat nor was it confined by textbook associations; whatever was useful was used.

Whilst on the Island I worked for Colin Tubbs of the Nature Conservancy Council and then moved across the Solent to the New Forest. As boss, mentor and lifelong friend, Colin gathered together a like-minded team to bring the recently assented Wildlife and Countryside Act into effect. In the early 1980s saltmarshes and their intertidal flats were particularly vulnerable to the growing affluence of Solent City. As well as the demands on the coast for industry, transport and urban growth there were burgeoning aspirations for leisure moorings, marinas, marina villages and all the paraphernalia that accompanies recreation. There was a constant need to resist attrition whilst pressing for better designations and a supportive policy framework. To compensate for the administrative grind of Town and Country Planning our work also gave us experience of those ways of farming that were sympathetic to the estuaries. There was so much to learn from people whose livelihoods supported wildlife outside nature reserves. It became clear that without such people and a supportive rural economy the wildlife we valued would fall into disrepair. There is a paradox in that saltmarshes can be beautiful in their dereliction and not wholly devoid of wildlife interest. Being able to compare the outcomes of different management regimes convinced me that an abandoned coastline deprived us of the rich diversity that comes with pastoral farming.

The invitation to write this book gave me the opportunity to bring the saltmarsh elements of my life to centre stage. By drawing on a selection of marshes from around Britain I have sought to explore themes relating to the wildlife of the marshes together with the relationship of people with those places. The selection of saltmarshes is personal and arbitrary but I hope to have included a fair cross section from across the three nations of Britain. Each chapter is written to stand alone as well as contributing to an unfolding story.

I am indebted to many friends and correspondents who have wittingly, or otherwise, helped me draw together the strands of thought that have inspired this book. The depth of knowledge of Britain's wildlife and wild places held in the collective memory of fellow naturalists never ceases to amaze me, nor does their open-handed generosity in sharing that wealth. The interpretation of what they have told me is mine alone, as are any errors or misunderstandings which may have crept in.

My special thanks go to:

Chris Archbold, Malcolm Ausden, Gemma Bodé, Jamie Boyle, Doreen Bruce, Andy Byfield, Dave Cadnam, Tim Callaway, Bob Chapman, Catherine Chatters, Eric Clement, Trevor Codlin, Andrew Colenutt, Jack Coughlan, Jonathan Cox, Dominique Cragg, Stuart Crooks, Bernie D'Arcy, Pete Durnell, John Durnell, Phil Dyke, Ian Evans, Andrew Excell, Steven Falk, Tim Ferrero, Ro Fitzgerald, Harry Green, Larry Griffin, Andy Harris, Tom Haynes, Sue Helm, George Hounsome, Pete Hughes, Richard Jefferson, Sorrel Jones, Graeme Kay, Sue Lawley, Peter Marren, Tim McGrath, Adam Murphy, George Parry, John Poland, Ian Ralphs, Martin Rand, Rebecca Read, Zoe Ringwood, David Robinson, Julian Roughton, Fred Rumsey, Neil Sanderson, Colin R Scott, Mike Smart, Mike Smith, Mark Spencer, Bill Waller, Brett Westwood, Debbie Whitfield and Eddie Wiseman.

My thanks also go to the many librarians and curators for their patience in retrieving stacked volumes and guiding me through their new technologies. I would particularly like to thank those who care for the special collections at the Hartley Library (University of Southampton), the National Oceanographic Centre (Southampton), the University of Winchester Library, the local studies collection at Gloucestershire Libraries, the Herbarium at the British Museum (Natural History), Comunn Eachdraidh Uibhist A Deas and the librarians at the Linnean Society of London.

I am indebted to the team at Bloomsbury for guiding me through the processes required to convert a naturalist's enthusiasms into a beautifully produced book. My thanks go to Brad Scott, copy-editor, to Susan McIntyre who gave order to the eclectic imagery, and especially to Katy Roper, commissioning editor, for her unfailing courtesy in curbing my digressions and helping me to focus on what really matters.

Introduction

S altmarshes occur wherever flowering plants grow in wetlands influenced by salt; this simple set of circumstances embraces an exceptional diversity of landscapes and habitats. Despite being an important part of the landscape of our island nation there is no common appreciation of what constitutes a saltmarsh or where those wetlands become something else. What follows is an exploration of early written accounts of saltmarshes together with an introduction to the key concepts used to describe these habitats and to understand their dynamic nature.

In plain English

Our ancestors recognised saltmarshes long before the arrival of ecologists and their concept of habitats. As a written word 'saltmarsh' has a pedigree predating the Norman conquests. The Anglo-Saxon Junius manuscript in the Bodleian library includes a poetic re-telling in Old English of the book of Exodus. In this heroic account the children of Israel are led safely across the *sealtne mersc* within which the pursuing Egyptians perish.

From the early medieval period onwards variations on the English word 'saltmarsh' appear with increasing frequency in documents as a landscape type, a place name and a surname. The equivalents in the Welsh language and Scottish Gaelic are similarly recorded as *Morfa hallt* and *Fìdeach*. The words have never fallen out of use nor changed their fundamental meaning of being wet salty places.

Not all early accounts referring to British saltmarshes were written in English. In the mid-13th century Walter of Henley wrote his treatise on husbandry in a French dialect. Lamond (1890) collated translations of various editions of his work which suggest that Walter

OPPOSITE PAGE:
Saltmarsh at Rodel,
Isle of Harris.

13

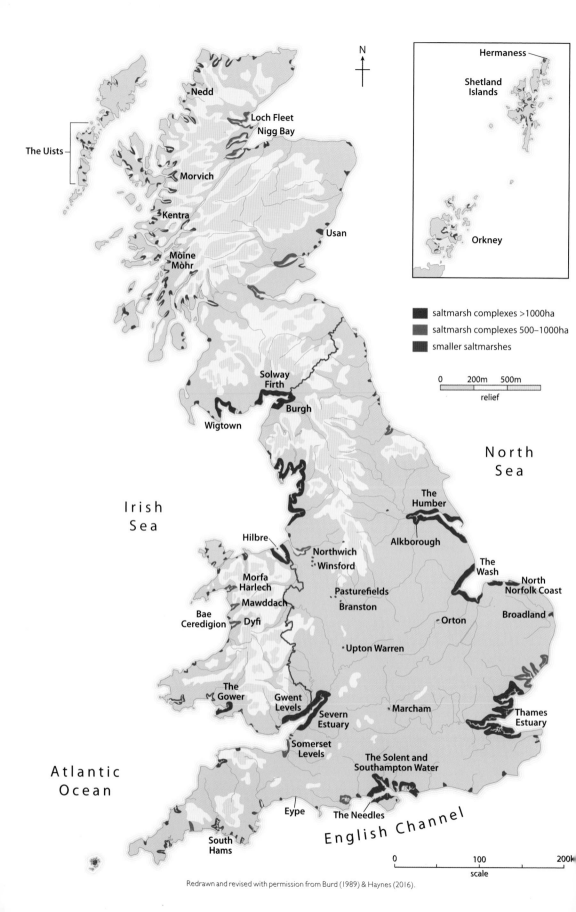

N

Hermaness
Shetland
Islands

Orkney

Nedd

Loch Fleet
Nigg Bay

The Uists

Morvich

Kentra

Usan

Mòine
Mòhr

■ saltmarsh complexes >1000ha
■ saltmarsh complexes 500–1000ha
■ smaller saltmarshes

0   200m   500m

relief

Solway
Firth

Burgh

Wigtown

North
Sea

Irish
Sea

The
Humber

Alkborough

Hilbre

Northwich
Winsford

The
Wash

North
Norfolk Coast

Morfa
Harlech

Pasturefields

Broadland

Mawddach

Branston

Orton

Bae
Ceredigion   Dyfi

Upton Warren

Atlantic
Ocean

The
Gower

Gwent
Levels

Marcham

Thames
Estuary

Severn
Estuary

Somerset
Levels

The Solent and
Southampton Water

Eype

The Needles

English Channel

South
Hams

0   100   200k

scale

Redrawn and revised with permission from Burd (1989) & Haynes (2016).

was probably an estate manager, possibly for properties owned by Canterbury Cathedral. Walter emphasised the importance of *mareys salyne* in providing particularly rich pastures. Saltmarshes supported two cows or twenty ewes in producing milk to manufacture a 'wey' of cheese and half a gallon of butter every week throughout the summer. He contrasted this abundance with grazing animals on wood pastures, mown meadows and stubble fields, which would require three cows or thirty ewes to produce similar quantities.

The earliest English definition of the word similarly derives from the French. Noel Chomel's *Dictionnaire oeconomique* of 1709 defined saltmarshes for their utility as 'a sort of grazing ground near the sea, which is commonly very rich land'. Originally published as a discourse on estate husbandry, the dictionary was translated into English in 1725 by Richard Bradley, the first Professor of Botany at Cambridge.

Time and science have not added clarity to the plain English word. Contemporary technical literature has generated a great many definitions of saltmarsh, each reflecting the perspective of the authors and their audience. In a review by Hough *et al.* (1999) the Environment Agency included a consideration of 22 definitions; the conclusion was that none of them provided a sufficiently rigorous description to allow the landward and seaward limits of a marsh to be accurately distinguished. Over a decade later Phelan *et al.* (2011) pragmatically recognised the fuzzy boundary between freshwater marshes and saltmarsh as being something impossible to define.

There is no universally accepted technical definition as to what comprises a saltmarsh. Within these caveats the national resource is currently estimated to be in the order of 48,000ha which is distributed around Britain's coast and in the English shires. Current best estimates presented by the government's advisors on the Natural Capital Committee (2015) are that this resource is declining in extent at about 100ha a year, that loss mostly accounted for by the rise in sea level outpacing saltmarsh development. As such, saltmarshes represent one of Britain's rarest and most threatened major habitat types, far scarcer than lowland peatlands, dwarf shrub heaths or sand dunes.

Key concepts

When exploring variations in saltmarshes and their dynamism it is helpful to understand key concepts that can be applied to every site.

OPPOSITE PAGE:
The distribution of saltmarshes in Britain: including selected locations.

Salinity

Water and salt are the very stuff of life; variation in the proportions of each ingredient determines what sort of life.

Salt in its purest form comprises equal quantities of atoms of sodium and chlorine combined into the compound sodium chloride. Seawater is predominantly diluted sodium chloride mixed with a wide range of other elements of which magnesium, sulphur, potassium and calcium are the most abundant. The concentration of dissolved salts in seawater around our coast is in the order of 35 parts per thousand. It may be easier to visualise this as 35g of table salt in a litre of water, or an unwholesome diet of 29 bags of crisps washed down by a small beer. The unit of 35 parts per thousand is a useful standard against which to compare wetlands of varying salinities, ranging from the mildly brackish through to the hyper-saline.

Hyper-saline conditions occur where saltwater becomes concentrated in solutions greater than the strength of seawater. In coastal marshes these circumstances are found in wetlands temporarily isolated from tidal flooding which experience rates of evaporation exceeding dilution. This happens when there is hot, dry weather between spring tides. Such habitats are rare, not least because Britain enjoys a relatively wet climate. For centuries the natural processes creating hyper-saline conditions have been exploited by industries extracting salt from seawater. The habitats created by these industrial processes can be equally supportive of the highly specialised organisms of hyper-saline conditions as their naturally occurring counterparts.

Brackish is a helpfully ambiguous word describing water not as salty as the sea. The degree and extent of dilution is determined by local circumstances and can change as quickly as the weather. As with hyper-saline habitats, a long history of coastal engineering has greatly altered the natural extent and distribution of brackish wetlands. Outside the Highlands there are few British saltmarsh landscapes where the full expression of such habitats survives intact. Coastal engineering has truncated the brackish zones of many tidal rivers and saltmarshes, but those self-same modifications have artificially expanded the fresher elements of the brackish zone, particularly in the ditches of coastal grazing marshes. The boundary between brackish and freshwater is fuzzy, with saltmarsh elements persisting in wetlands where salt is diluted to less than two parts per thousand. Such mildly brackish dilutions remain detectable to the human palate; the conditions are piquant but not indigestible.

As a general rule saltmarshes accumulate species richness and diversity as they become less salty. Without apology I have sought to seek out and celebrate the most intact of Britain's saltmarshes, not least to encourage a wider appreciation that the habitat is a great deal more diverse than what lies between average tides.

Tides

All habitats are dynamic; saltmarshes are particularly so in readily comprehensible timescales. The development of a saltmarsh is driven by processes ranging from the daily to the millennial.

The overwhelming majority of British saltmarshes are tidal. Twice daily a whaleback of water makes its westward passage around the world. The bulge in the oceans created by the gravitational pull of the moon and sun is expressed as the highs and lows of the tide as it passes around our coastline.

In theory tidal ranges are predictable. Over the course of a lunar month there is a fortnightly cycle of spring tides when the tidal range is at its greatest, together with neap tides when the range is at a minimum. Spring tides occur shortly after the full and new moons, with neap tides equidistant between them. The lowest neaps and the highest springs occur at the spring and autumn equinoxes when the gravitational pull of the sun is at its greatest.

The highly indented coast of Taynish and Loch Sween, Argyll.

An extreme high tide, Brancaster, Norfolk.

The frequency of tides is determined by astronomical cycles but the height of each tide varies depending on the character of the coast. The tidal range of the open coast in Britain is in the order of 3m. In funnel-shaped estuaries, such as the Severn, local geography can concentrate the volume of the water to regularly generate tides in excess of 12m. On highly indented coastlines, such as around Loch Sween, the converse may be the case, the spring tide here being limited to little over a metre. Each saltmarsh will reflect the particular characteristics of its local tidal regime.

Predictable tides are subject to unpredictable variations in the weather. A low-pressure weather system will result in higher tides as the atmospheric pressure on the ocean surface is reduced. Low-pressure systems are associated with storm events. A strong wind can magnify the effects of the tide by driving water before it, raising the height of the sea and restricting the ebb flow. If the storm persists there is a cumulative effect with following high tides adding to the mass of water. Such extreme storm-surge events pose catastrophic risks to human life and property. Conversely, a high-pressure weather system will suppress the tidal range, leaving the higher reaches of a marsh vulnerable to desiccation as high-pressure systems are associated with sunny, droughty conditions. South of the Highland line most major saltmarsh ecosystems in Britain have been modified to restrict the reach of high tides. Through the erection of sluices and seawalls and by embanking rivers we have wrested land from the intertidal. In recent centuries the availability of technology and capital has enabled such modifications to be intensified to a degree that extensive areas of former saltmarshes have been transformed into something entirely different.

Relative sea levels

The concept of relative sea level is less easy to grasp than salinity, which you can taste, and tides which you can see. Sea level is something that changes with the climate over geological time; its effects are incremental but not so subtle that they cannot be measured within the span of a human lifetime.

Changes in the climate drive changes in solid geology. During each of the glaciations over the last 2.6 million years the immense weight of ice has compressed and displaced the crust of the earth. Each time the ice melted, pressure was lifted and the solid geology of what is now Britain continues to experience the slow process of rebounding. The latest phase of the current cycle of glaciations resulted in ice caps across the north of Britain concentrated on what are now the Highlands of Scotland. For a little over 12,000 years the Highlands have been rising whilst adjoining areas are subsiding. The Outer Hebrides, Orkney and the Shetlands are in the zone of subsidence, as are those parts of England and Wales south of a line from the River Tees to the Llŷn Peninsula. This process is known as isostatic readjustment. Readjustment is not as neat as a geological see-saw, with the centre rising as the periphery descends; the overview is sound but the local details are more subtle. In place of a see-saw another analogy may be an old leather sofa, relieved of its sitters, slowly returning to some semblance of its former self.

The central belt of Scotland is currently rising at a little under 1mm per year with the East Anglian coast subsiding to the same degree. The process of uplift in the north and north-west means that there are shorelines which were formed thousands of years ago which today can be found raised well above the tide. The Scottish coast has many such raised beaches and 'stepped' saltmarshes. In these circumstances saltmarshes persist over time where they continuously colonise the lower shore as it is lifted out of the subtidal into the intertidal. To survive in such circumstances the marsh must constantly grow seaward.

Invernaver, raised beach and tidal river, Sutherland.

Oaks killed by the rising tide, South Hams, Devon.

The opposite is the case in southern England where historic shorelines are being drowned and becoming progressively subtidal. Indeed, archaeologists wishing to study first-hand how prehistoric people lived in coastal landscapes need to learn how to dive. To persist in such circumstances a saltmarsh must either migrate inland or grow vertically by accreting sediments. To stand still the migration and growth of a southern saltmarsh needs to be at the same pace as the land beneath it subsides.

These shifts in solid geology are complemented by the effects of climate change on the oceans. Current sea level is a reflection of the proportion of the world's water held as ice; as the world warms so ice fields melt and sea levels rise. This rise occurs not only because there is more liquid water but also because that water is warmer; the warmer the oceans then the greater their volume through the physics of thermal expansion. These natural phenomena are being supplemented by people through the effects of carbon emissions from the burning of fossil fuels.

The practical consequences of these processes are changes in the level of the sea relative to the land. This is particularly important since many cities, such as the capitals Cardiff, Edinburgh and London, all sit in coastal locations. The current best estimates of relative sea-level rise between 1990–2095 identify Edinburgh as the least vulnerable of the three capitals to tidal flooding.

All life on the coast, be that human or wildlife, needs to adapt to these changes if it is to survive.

Estimates of relative sea-level rise: 1990–2095

	Cardiff		Edinburgh		London	
	High CO_2 emission scenario	Low CO_2 emission scenario	High CO_2 emission scenario	Low CO_2 emission scenario	High CO_2 emission scenario	Low CO_2 emission scenario
1990–2020	11.5cm	8.2cm	7.5cm	4.3cm	11.5cm	8.2cm
1990–2040	20.8cm	14.8cm	14.2cm	8.2cm	20.8cm	14.8cm
1990–2060	31.4cm	22.2cm	22.1cm	13.0cm	31.4cm	22.2cm
1990–2080	43.3cm	30.5cm	31.4cm	18.6cm	43.3cm	30.5cm
1990–2095	53.1cm	37.3cm	39.2cm	23.4cm	53.1cm	37.3cm

Source: adapted from Millin 2010 © UKCIP, 2010.

Changes in relative sea-level are not new, and over geological time the pace of change has varied. In the modern age people have grown to rely on coastlines being stable; we have invested in the shoreline remaining where we mistakenly perceive it has always been.

Latitude and climate

In the far north of Britain the Shetlands share their latitude of 63°N with southern Greenland, the Canadian sub-Arctic and the ice-bound straits between Alaska and Siberia. In contrast the south of Britain lies at 50°N on a parallel with Frankfurt and the Altai mountains of central Asia. Our climate is unlike any of these equivalent global regions due to the moderating influence of the Atlantic Ocean in general and of the warming effects of the Gulf Stream in particular.

The combined effect of latitude and oceanic influences means Britain's vegetation includes elements of floras associated with diverse climatic zones. There are arctic-alpines growing in Britain's saltmarshes, particularly in the north, together with circumpolar species known from around the sub-arctic. Dilution of saline influences in a temperate continental climate brings with it conditions similar to the Baltic, northern Europe's mediterranean sea. There is a suite of British species associated with such brackish conditions, particularly in the higher rainfall areas of the north and west. The epithet 'Mediterranean' is more usually used to describe the sea between Africa and Europe, a place of much lower latitudes than Britain with a distinct climate of hot summers and mild winters. Species with a predominantly Mediterranean distribution are found in British saltmarshes particularly on its southern and eastern shores.

Latitude and climate help to explain the overall pattern of the distribution of Britain's vegetation. In practice, the distribution of any species is a reflection of its individual environmental tolerances combined with the availability of suitable habitat, together with chance.

Herbivory

Wherever there are plants there will be creatures feeding on them as all vegetation is naturally subject to herbivory. Britain's saltmarshes share a common evolutionary history in that their ecosystems developed over many interglacials in the presence of a range of herbivores. In the current interglacial most of the native large herbivores of our coastal wetlands have been displaced by domesticated animals.

Derek Yalden's collation of archaeological evidence (1999) indicates that less than four thousand years ago, Aurochs, Wild Cattle *Bos primigenius* and Elk *Alces alces*, along with Red and Roe Deer *Cervus elaphus* and *Capreolus capreolus*, were native in Britain. Mammals were not the only significant grazers of coastal wetlands; waterfowl, particularly flocks of swans, geese and grazing duck, would have also had a substantial effect in modifying vegetation.

There is a lively debate as to what degree waterfowl are dependent on large mammals opening up grasslands or whether they are capable of sustaining their grazing grounds independently. In a natural state it is likely both groups would have simultaneously exploited the same

People lived alongside Aurochs for thousands of years.

resources. In contrast herbivorous small mammals and invertebrates are only able to modify vegetation on a localised scale; they rely on the larger creatures to create the vegetation structure and habitat niches on which they depend.

Over time the largest of the wild herbivores were progressively displaced by much smaller domestic cattle with non-native sheep playing an increasingly important role as a source of food, fibre and fertility. The evidence for Wild Horses *Equus ferus* surviving in Britain through the immediate post-glacial period in any numbers is slight, but by the Bronze Age domesticated horses regularly appear in the archaeological record. For those species whose saltmarsh niches are sustained by bovine behaviour a domesticated cow is likely to be an effective substitute for an Auroch. Species dependent upon modifications to the landscape arising from the influence of Elk or the testosterone-fuelled antics of bachelor Aurochs would be less well served.

Domesticated livestock are the ecological successors of our native mammal fauna.

The ecological effects of grazing animals on saltmarshes are similar to other open habitats. The presence of large herbivores suppresses the dominating effects of coarse perennial species and diversifies the structure of vegetation. Herds and flocks create areas with short swards and bare ground, opening up niches for colonisation whilst

Organic strandline debris.

Coenosia karli, a muscid fly
of saltmarsh strandlines.

redistributing nutrients through behavioural traits such as dunging, urination and death. The particular circumstances of animals and birds grazing over intertidal areas mean their behaviour is influenced by access to freshwater and refuge from tidal floods. The tide itself redistributes nutrients; a sheltered saltmarsh bay can sport a strandline of organic debris including the dung of grazing animals, and such natural concentrations of fertility sustain one of the rarest of all natural habitats, the truly eutrophic.

Over the last century many British saltmarshes have undergone profound changes where domestic livestock have been withdrawn but not replaced by their wild progenitors or some suitable substitute. Unlike fully terrestrial habitats, British saltmarshes don't tend to scrub over and develop into secondary woods. The decline in quality and diversity is masked under dwarf shrubs, usually Sea-purslane *Atriplex portulacoides*, together with Common Reed *Phragmites australis* and other coarse grasses. The resulting habitats may be beautiful but they are incomplete as the absence of large animals diminishes the full expression of the diversity of the marsh.

Halophytes

Ecological literature relating to saltmarshes often refers to the term 'halophyte', which derives from the Greek *halo* = salt, *phyte* = plant. This emerged in the European literature of the 19th century and was adopted in England by the 1880s, but the ideas it seeks to represent can be traced to Johann Wolfgang von Goethe's book of 1790 *The Metamorphosis of Plants*. Goethe recognised that plants from many different families have a tendency to share similar characteristics where they are growing in similar environments. He is remembered today more for his dramatic rendition of Faust selling his soul to the devil than for his work as a botanist and anatomist.

Goethe's observations were taken up by Eugenius Warming (1841–1924) of the University of Copenhagen. The century between Goethe and Warming was one of exploration by Europeans of, what was for the explorers, the remote regions of the world. Drawing on experience from across temperate and tropical climes, Warming described common features in plants and plant communities reflecting responses to the environmental stresses under which they had developed. Very different species from distant continents could be found with similar life forms and in similar communities. His studies of saltmarshes demonstrated the shared characteristics of a selection of plants growing under the stresses of salinity, saturated soils and the pull of the tides. These species have a tendency to be well rooted, succulent and adapted to flowering and setting seed between saline submersions.

The characteristic features of a halophyte are evident in the complex suite of species which make up the glassworts, *Salicornia* and *Sarcocornia* spp. In these plants the leaves and flowers are reduced to mere scales to reduce the area exposed to drying out by sun and salt water. The surface layers of such species are thick and waxy to resist desiccation and guard internal freshwater reserves. Pores essential

Glassworts grow to maturity in a single short season.

for respiration are reduced to a minimum, with some adapted to secrete salt from the body of the plant; these and other features of halophytes are shared with species adapted to desert conditions. Highly specialised organisms such as glassworts are dominant components of some saltmarshes but are certainly not ubiquitous across all saltmarsh types.

Despite the longevity and frequency of the use of the word 'halophyte', its definition remains equivocal. Dependence on, or tolerance of, salts is found to a degree in all plants, not least as salts are essential plant nutrients. Attempts have been made in many learned works to find a definition. Some of the earlier attempts sought to express this by the quantity of salt present. There was however no consensus as to how much salt there should be and for how long it should be present before a plant could be considered a halophyte. More recent attempts at definitions have drawn on the physiology of plants, with practical definitions emerging to assist progress in growing food in arid lands.

For all practical purposes 'halophyte' remains a loosely defined term of convenience. Not all species of saltmarshes are halophytes and there are many apparent halophytes that will thrive under alternative stresses in the absence of salt. There is a limited number of organisms whose distribution is wholly confined to saltmarshes; it is a habitat of many generalists and few specialists.

Early ecological descriptions

It is easier to describe a habitat than to define it. Saltmarshes have attracted ecologists since the foundation of the science of ecology. Those early studies still exert a profound influence on how saltmarshes are popularly perceived today.

The *Journal of Ecology* was the first academic journal in the world dedicated to the science of ecology. First published in 1913, the first scientific paper in the first volume was a description of the saltmarshes and shingle spit of Blakeney Point in Norfolk. The author was Francis Oliver and the editor was A G, later to be Sir Arthur, Tansley. Between 1911 and 1939 Oliver and Tansley, together with their Cambridge colleague Val Chapman, published a series of ecological accounts of Norfolk's saltmarshes. The marshes they describe support strikingly clear zones of vegetation reflecting the frequency of inundation by the tide. Through the decades their descriptions became more detailed,

with Tansley's 'general saltmarsh' of 1911 being reworked to reflect a range of communities.

By the 1930s saltmarshes of the North Norfolk Coast were described by Tansley as comprising three broad communities, lower, middle and upper marsh. The lower marsh was defined as being subject to 50 hours or more of submergence in each of the summer months. This was the zone characterised by eelgrasses *Zostera* spp. and glassworts. The middle marsh was defined by elevation, running from 3.5 feet below Ordnance Datum, being the point of 50 hours of submergence, to 1.3 feet above. This is equivalent to 1m below and 40cm above Ordnance Datum. The middle marsh is characterised by Sea Aster *Aster tripolium* with other perennials including Sea-purslane and Sea Plantain *Plantago maritima*. The upper marsh is in the zone between the middle marsh at 3.5 feet above Ordnance Datum to a point where monthly submersion is limited to just three hours. The upper marsh is above the zone of Sea Aster and whilst retaining many of the species of the middle marsh it is typified by the presence of Sea Rush *Juncus maritimus*. The upper marshes of the North Norfolk Coast with their transitions into shingle are the national stronghold of Shrubby Sea-blite *Suaeda vera* and Sea-heath *Frankenia laevis* which are woody perennials with an affinity to the Mediterranean coast.

Arthur and Edith Tansley visiting the University of Chicago, 1913.

27

Sea-milkwort blooming beneath a canopy of Mediterranean Matted Sea-lavender *Limonium bellidifolium*, Blakeney.

In these early studies there was some appreciation of regional differences. The marshes of Ynyslas on the Dyfi were recognised as having five plant associations in just two communities; the upper and lower marsh. The Ynyslas marsh descriptions reflect the preponderance of grasses rather than herbaceous perennials with associations characterised by Common Saltmarsh-grass *Puccinellia maritima* and Red Fescue *Festuca rubra*.

By monitoring Norfolk's marshes over decades, Tansley and his colleagues were able to demonstrate the effectiveness of the species of the lower marsh in stabilising and accumulating silt, so enabling the establishment of middle and upper marshes. Their studies were timely in witnessing the rapid colonisation of mudflats by the newly-described Townsend's Cord-grass *Spartina townsendii*. Tansley observed in 1939: 'No other species of salt-marsh plant, in north-western Europe at least, has anything like so rapid and great an influence in gaining land from the sea'.

Tansley's focus however remained on the middle zones of the marsh. In the 900 pages of his 1939 account of *The British Isles and their Vegetation* the descriptions of maritime communities take up 88 pages, of which only one addresses brackish conditions. To his credit, Tansley concluded that his studies of these habitats had been 'far from exhaustive, and that a number of different, mainly unexplored, communities are represented'.

Whilst magnificent and readily accessible to academics, the marshes of the North Norfolk Coast are no more typical of British saltmarshes than anywhere else. The character of the Norfolk marshes reflects their particular circumstances, not least the range of sediments available for colonisation together with the scarcity of substantial freshwater inputs. The legacy of these classic studies still influences how ecologists learn about saltmarshes. It is left for speculation how we would regard the habitat today if the home of saltmarsh ecology had been Inverness or Aberystwyth.

Vegetation traps silt across the tidal range.

Contemporary descriptions

Coastal saltmarshes are conventionally described through a combination of their geomorphology and vegetation.

The marshes that Tansley and his colleagues studied on the North Norfolk Coast are classic barrier-connected marshes. Sediments have collected and become vegetated in the lee of the shingle spits and sand dunes of Blakeney Point and Scolt Head Island.

Foreland marshes do not benefit from such barriers and are relatively exposed to the open sea. The exposure may be ameliorated by being in a bay, such as the Wash, or benefiting from submerged offshore banks. Such a marsh has formed on the Denge Peninsula in Essex where the marshes of the Thames open out into the English Channel.

Estuarine marshes develop in sheltered tidal waters where there is an appreciable flow of freshwater. Estuaries can be found on various scales from a discrete few hectares of a West Country ria to the extensive coastal wetlands of the Thames, Dyfi or Solway. The presence of freshwater brings with it opportunities for marshes to exhibit transitions from seawater to a wholly fresh environment.

In contrast to our neighbours in the low countries of Europe a significant proportion of Britain's coastline is rocky. The long inlets of sea lochs provide shelter for saltmarsh formation. In the shallower sea lochs saltmarshes form in bays, at the outfall of rivers and in the intertidal alongside the loch edge.

Saltmarsh

ABOVE: Brackish estuarine marsh, Beaulieu River, Hampshire.

BELOW: A loch head saltmarsh delta, Loch Ainort, Isle of Skye.

A great deal of research has been undertaken into the formation of creeks and pools within saltmarshes. Creeks can be highly distinctive, the dendritic alveoli of the quiet backwaters of East Anglia being entirely different from the linear trenches reflecting the high tidal energy of the Severn. Creek edges provide more freely drained soils than the surrounding marsh and can be picked out through the presence of distinctive communities. Saltmarsh pools vary in character with great variability in vegetation, salinity and consistency in retaining water. Pools, whether they are vegetated or not, support some of the most highly specialised species of intertidal marshes. Despite the diversity of their form, the shape and origin of creeks and marsh pools has no significant effect on the species that inhabit them.

Scoured trenches on a Severn marsh, Somerset.

The early 20th-century descriptions of British saltmarsh vegetation by Tansley and his colleagues were substantially revised in the 1970s by Paul Adam. Adam took over 3,000 samples of saltmarsh vegetation from across Britain and laid the foundation for the National Vegetation Classification of British Plant Communities.

Saltmarshes were specifically addressed in the fifth and final volume of *British Plant Communities* in 2000. In his introduction John Rodwell paid tribute to the insight of Tansley and his colleagues. Whilst venerated, these classic descriptions required updating. The

Marsh pools flushed with freshwater, Landimore, the Gower.

sampling that informed the National Vegetation Classification sought to describe the character of broadly similar areas of vegetation throughout Britain rather than focus on the interesting or unusual. The classification therefore does not describe exceptional sites such as brackish woodlands and the full range of perched saltmarshes. The objective was to provide a vocabulary to describe characteristic communities, and how they relate to one another, together with the environmental variables that influence their composition and distribution. The breadth of the sampling was recognised to have its limitations, particularly in the vegetation of brackish open water and ditches. These less saline elements had fortunately been partially covered in previous volumes. Setting those caveats aside, for the first time there was a readily accessible national overview of the character of British saltmarshes. The resulting classification challenged traditional perspectives and covered a suite of habitats described by John Rodwell as 'far from the common conception of salt-marsh vegetation'.

The National Vegetation Classification describes 44 plant communities from saltmarshes, ranging from the vegetated sub-tidal to brackish wetlands. The subtleties reflected in these descriptions resist further compression. Since its publication in 2000 there have been proposals to add to its accounts but nothing has sought to replace them. The monumental achievement of the National Vegetation Classification requires the serious student to consult the original work in its entirety.

Saltmarsh plant communities reflect environmental variables ranging from global processes to local circumstances; the following accounts from across Britain explore that diversity.

A summary of saltmarsh communities described by the National Vegetation Classification (NVC)

Broad habitat type	Number of NVC communities	NVC reference	Notes
Eelgrass and Tasselweed communities	3	SM1–3	These communities tend to be in the intertidal and sub-tidal but may also be present within the shorter SM communities and exceptionally in pools landward of coast defences.
Lower saltmarsh	25	SM4–28	The SM communities are arranged so that they reflect the pioneer communities of the lower marsh through to the highest reaches of the tide and strandlines. Broadly, SM4– SM14 are lower marsh, SM13–SM23 middle marsh and the remainder upper marsh and ephemeral communities. There is considerable overlap in communities in their distribution, with some communities found across the tidal range where circumstances permit.
Middle saltmarsh			
Upper saltmarsh			
Brackish flood pastures and mire	4	MG11–13, M28	These mesotrophic grassland, mire and swamp communities are found in transitions to freshwater habitats. The communities reflect environmental and management variables, particularly salinity and grazing pressures.
Brackish swamps	7	S4, S12, S18–21, S28	
Saltmarsh strandlines	2	S26, OV24	The strandline communities reflect the nutrient-rich zones at the base of seawalls and at the extreme tidal limits of coastal rivers.
Brackish open water	3 (+1)	A6, A12, A21 (+ Type J)	These communities are found in open water within brackish habitats. Further work is required to describe them. In 2006 a classification of British lakes by Duigan et al. following NVC methodology identified the communities of brackish lakes as Type J.

Source: Rodwell 1991, 1992, 1995 & 2000. Adapted from *British Plant Communities* edited by John Rodwell, 1991–2000, © NERC © NCC 1991, © JNCC 1992–2000, used with permission from the Joint Nature Conservation Committee.

Far from the shore

A worldwide distribution map of saltmarshes will show concentrations of the habitat in two distinct zones, the coastal and the continental. The geography and climate of Britain are unsuitable for the formation of salt lakes and other features of continental saltmarshes. However, the underlying geology of central England includes salt-bearing rocks that support one of Britain's most intriguing wetland habitats.

Pasturefields

At first sight the Staffordshire Wildlife Trust's nature reserve at Pasturefields is unremarkable. Its fields are bound on one side by the Trent and Mersey Canal running alongside the busy A51. The floodplain rises gently towards the canal and the canal flushes water back to the Trent. Pasturefields escaped the worst excesses of agricultural intensification but its grasslands are far from pristine. A study in the early 1970s reported one field, at least in part, ploughed in 1960. The presence of ridge and furrow over the higher ground testifies to earlier cultivations. These are relatively fertile alluvial soils annually enriched by flooding; there are neither softly romantic floral displays nor drifts of guarded rarities.

A mottled matrix of greys and greens defines the turf of the valley grassland; there are rushes and plantains and arrowgrass, as one may expect from such a spot. It is when one sets about appreciating the details that the extraordinary nature of the place is revealed. The plantain is Sea Plantain *Plantago maritima*, the rushes are Saltmarsh Rush *Juncus gerardii*, the arrowgrass is Sea Arrowgrass *Triglochin maritimum*. In the more closely grazed and poached muddy hollows

OPPOSITE PAGE:
Pride of Flash Pool.

35

Pasturefields saltmarsh
from the canal.

are flowers of Lesser Sea-spurrey *Spergularia marina* and Sea-milkwort *Glaux maritima*. Both vernacular and scientific names spell it out; this is a saltmarsh, a place of seaside specifics, *marina*, *maritimum*, *maritima*.

In the late 1680s, Dr Robert Plot, Professor of Chemistry at Oxford, was given hospitality and support by Walter Chetwynd in preparing his 1686 *Natural History of Staffordshire*. Chetwynd was the owner of Ingestre Park whose grounds adjoined what is now the nature reserve. Plot was fascinated by springs, particularly salt springs as they supported his theory of linkages to the sea by means of subterranean passages.

Chetwynd and Plot were both Fellows of the Royal Society and so would have been familiar with the work of their contemporary, Robert Hooke, the Society's Curator of Experiments and promoter of microscopes. The brine from the springs on Chetwynd's estate when viewed under that 'ingenious contrivance' was found to contain 'a great multitude of very minute animals swimming about in it'. Plot was unable to offer a name to these creatures beyond describing them as insects.

Walter Chetwynd had a personal chaplain, Charles King, who was also a botanist. King showed Plot specimens of Sea Aster *Aster tripolium*, growing at Ingestre, which he called Sea Starr-wort, 'near the place where the brine of itself breaks out above ground, frets away the grass, and makes a plash of salt-water'. Plot reflected on the presence

Sea Aster has been lost from the saltmarshes around Ingestre.

of this seaside plant so far inland 'which though generally said to grow upon the sea-coasts, especially in saltmarshes, where the tide ebbeth and flowest ... yet here it is found in an inland country, at least 50 miles from the sea'. He concluded the occurrence was natural: 'it seems not at all to have been out of its natural abode'.

The scale and vigour of the salt springs at Ingestre equally impressed Plot where they fed saltworks as where they remained in a natural state. He was not immune to aesthetic considerations and observed, 'The subterraneous brine is so strong that the cattle standing in it in summertime and throwing it on their backs with their tails, the sun so candies it upon them that they appear as if covered with hoar frost'.

Over a century later in 1817 the owners of the neighbouring manor of Tixall produced 'Flora Tixalliana', a description of wild plants within a short ride of their home. Thomas Clifford and his brother Arthur published their natural history in Paris, an interesting choice for the time, and did not feel constrained in commenting on the management of their neighbour's estate. 'Wandering down the Vale of Trent, the eye is attached by the singular appearance of volumes of white smoke, perpetually rising from the salt works of Shirleywich, over the lower groves of Ingestrie'.

The salt industry was clearly having an impact on wildlife. In their account of Sea-milkwort, which they called Black Saltwort, they observed 'The immense drains made by Lord Talbot [the owner of

Ingestre and the saltworks] in the adjoining saltmarsh of Ingestrie, for the purpose of carrying off the salt spring, may probably destroy this and other marine plants'. The impact of drainage on wading birds was also noted: 'So much land has been drained in this parish of late years, that there is hardly any to be found of that wet spongy ground, which they chiefly delight'.

The brothers recognised the unusual nature of inland saltmarshes. In describing Sea Club-rush *Bolboschoenus maritimus* at Shirleywich their account includes the comment that the plant is 'in different saltmarshes in the neighbourhood of Tixall, as are to be found, this and other marine plants which have been generally supposed to grow only on the sea coast'. Amongst the saltmarsh species they also recorded Sea Aster, using the contemporary name of Sea Starwort, which they describe growing in a saltmeadow near Shirleywich, between the Trent and the canal, a location remarkably similar to Pasturefields.

With the growing popularity of natural history through the 19th century many other observers added to the records of saltmarsh plants. The flora of this short stretch of the Trent was extended to include Grey Club-rush *Schoenoplectus tabernaemontani*, Sea Club-rush, Distant Sedge *Carex distans*, Wild Celery *Apium graveolens* and Dittander *Lepidium latifolium*.

What survives today at Pasturefields is a remnant of a much larger, more complex series of saltmarshes. The area is still capable of producing surprises. The Saltmarsh Rush site at Ingestre Park's Lion Lodge was resurveyed as it fell within an early option for the route of a high-speed railway. The surveyor, Dave Cadnam, was challenged by what appeared to be an odd flower-spike of a saltmarsh-grass. On detailed examination the specimen proved to be Borrer's Saltmarsh-grass *Puccinellia fasciculata*, a species never previously known from the Trent.

England's landlocked saltmarshes

Pasturefields is an example of a continental form of saltmarsh. In Britain such marshes are confined to England where the majority are associated with springs arising from strata laid down under desert conditions in the Triassic period some 200–250 million years ago. These rocks are from the Mercia Mudstone Group and comprise a complex of mudstones, marly clays and salt-rich haline deposits. The character of the geology varies from the Severn Estuary to the Tees.

In places the salt-rich beds are thick, dry deposits such as at Winsford in Cheshire. Elsewhere groundwater has penetrated to produce subterranean brine streams with saltmarshes forming where springs break out on the surface. The collapse and flooding of cavities created through dissolving rock salt creates lakes locally known as flashes. Some of these flashes remain connected by springs to the underlying saline rocks. A provisional inventory of inland saltmarshes is set out in Appendix A. Each marsh has its own character; a journey through central England reveals something of that diversity.

Oxfordshire

The village of Marcham, north of Wantage, derives its name from the Old English elements *merece* and *ham*. The Oxford English Dictionary gives the derivation of *merece* as the plant *smallage*, which was known to medieval herbalists under a range of names including *anglice merce* and *anglice smalache*. Over time this pungent member of the carrot family was brought into cultivation and bred into something more palatable. By the late 17th century it gained a name familiar to us; smallage became celery. Cole *et al.* (2000) offer a translation of the Anglo-Saxon place name of Marcham as the meadow where the celery grows. Wild Celery *Apium graveolens* is a regular component of the higher reaches of saltmarshes, often growing on the margins of ditches and runnels.

Wild Celery.

George Claridge Druce's *Flora of Berkshire* (1897) included records from a salt spring near the village. At that time Marcham fell within the administrative county of Berkshire and remains so for the purposes of botanical recording. The flora described by Druce included the familiar components of other inland saltmarshes including Wild Celery, Saltmarsh Rush, Distant Sedge and Sea Arrowgrass, as well as the scarcer Parsley Water-dropwort *Oenanthe lachenalii*.

Regrettably, the marsh around the salt spring fell to the drive for agricultural improvement in the 1960s and by the early 1970s it was regarded as lost to continuous cereal cultivation. Local botanists continued to monitor the site and in the 1990s were rewarded by the rediscovery of Wild Celery germinating in a winter fallow following a breakdown in the land drains. A sympathetic farming regime was adopted, which has supported the Celery population in expanding to hundreds of plants. The Celery of Marcham is quite rightly a source of local celebration and pride.

Celery may not be the only species of Druce's marsh to have survived. There are teasing records from the 1990s of other species of the saltmarsh edge including Brookweed *Samolus valerandi* and Strawberry Clover *Trifolium fragiferum*, both of which may yet reappear. Recent surveys have confirmed the presence of Distant Sedge and have identified a putative population of Saltmarsh Rush. The saltmarsh flora of Marcham, at least in part, appears to have persisted in the seedbank. With the continuation of sympathetic management we can look forward to fresh surprises.

Gloucestershire

As recently as 1948 the Reverend Riddelsdell *et al.* described Wild Celery as growing 'throughout the Vale (where salt springs are frequent)', the Vale in question being the Vale of Gloucester. Previously the presence of such springs attracted attention not so much for their natural history as for economic opportunities. In his 1779 description of Sandhurst, north of Gloucester, the antiquarian Samuel Rudder wrote,

> *there is a spring of medicinal, or purging water, like the Cheltenham spaw, but it is sometimes dry in summer. And there is also a salt-spring, upon the discovery of which, about twenty years since, a pit was opened to considerable depth, with an intention of erecting a salt-work, but not proving so strong as was expected, that project was then dropt, and the mouth of the pit stopt up with bricks to prevent accidents.*

The concentration of both salt and purging springs around Redmarley D'Abitot in the north of the county also attracted Rudder's attention. He described a spring at Pauntley, 'of strong brackish taste, and very strong purging quality, rises out of a swampy place by the side of the Leadon and flocks of pigeons resort there to eat the salt made by evaporation from the water'.

In making this observation Rudder retells the folk story of pigeons assisting the discovery of springs upon which the successful

Cheltenham spa was developed. If a spa was subsequently promoted at Pauntley then the idea was a short-lived speculation. Nonetheless, the springs and an inscription 'Pauntley Spa (disused)' appear on the Ordnance Survey maps of the late 19th century in the wooded riverside of the Leadon.

Over the summer of 1913 John 'Jack' Wilton Haines explored the salt springs of the valleys of the Leadon and Ell around Redmarley D'Abitot including those at Pauntley, Ketford, Payford Bridge and Pool Hill. A solicitor by profession, Jack Haines was also a poet and botanist. Over his long life he was the confidant of many of the prominent artistic figures of his day, including Edward Thomas, a superior poet but lesser botanist, who lived for a while near Dymock, close to Pool Hill.

Not only did Jack Haines record wild plants, he was also a vigorous collector and preparer of herbarium sheets. The sole British record of Sea Beet *Beta vulgaris* ssp *maritima* from an inland saltmarsh, at Ketford, is supported by his pressed specimens of flower spikes now held in the Gloucester Museum. The cluster of salt springs along the Ell and Leadon was described by Haines as supporting Celery, Brookweed, Saltmarsh Rush, Sea Arrowgrass, Sea Club-rush, Grey Club-rush and Distant Sedge. Unfortunately, these springs appear to have been overtaken by the drainage and pasture management of conventional farming, with only fragments remaining of their former glory.

Sea Club-rush.

Worcestershire

It is a pleasant walk from the valleys of the Ell and Leadon to the Malvern Hills. Where the commons run along the lowland margins of the Malverns there are enigmatic populations of Slender Hare's-ear *Bupleurum tenuissimum*. This diminutive member of the carrot family has a distribution throughout Europe reflecting short turf on upper saltmarshes. There are historical records of Slender Hare's-ear from the lowlands of the Malverns and the contiguous roadside commons running from the hills to the River Severn at Ryhdd. Where cattle-grazing persists the plant can still be found, such as on Hollybush Common. Here, in an entirely atypical habitat, Slender Hare's-ear completes its annual life cycle on the sunny side of anthills in closely grazed seasonally parched slopes. There are no suggestions of any salt influence in the soils of Hollybush Common nor are any other saltmarsh species present. What makes conditions here suitable for Slender Hare's-ear is unknown; the stresses at Malvern that maintain the open conditions are severe but they are not saline.

In the early 19th century Slender Hare's-ear was also known from just over the river from Rhydd at Defford Common. Defford attracted

Slender Hare's-ear grows on the parched anthills of Hollybush Common.

botanists through the late 18th and early 19th centuries. It is possible the Slender Hare's-ear recorded there grew in a saltmarsh, or parched grassland, or possibly in both, as a range of habitats was present. The common also supported Lesser Sea-spurrey and Sea Club-rush, species of the margins and open water of salt wetlands. Amphlett & Rea (1909) collated historical records from the county including that of 1787 when Brookweed was described as growing from the side of a brook running from the brine pit at Defford Common, and in 1845 Parsley Water-dropwort was found near the saline spring. A salt spring is clearly marked at Defford on the Ordnance Survey maps of the late 19th century along with the adjacent wooded Salt Baths Covert, an extension to the neighbouring Croome Park. Defford Common is now unrecognisable as the place described in the 19th century. Following enclosure, what was the common became RAF Defford, and after the closure of the airfield in 1957 the land is now managed for farming and industry. The salt springs and brook persist as shadowy field marks in an expanse of arable.

Some way to the north of Defford is Droitwich, a town that grew up around its salt springs and was named by its Roman occupiers as Salinae, literally translating as 'the saltworks'. The natural springs were vigorous with pure and highly concentrated brine. Depending on the enthusiasm of the recorder, historical accounts suggest the springs were at least ten times, possibly twenty times, the concentration of seawater. It is reasonable to assume that saltmarshes naturally formed around these springs but they have been lost to the industrial processes of at least two thousand years of salt production.

Upton Warren lies a little to the north of Droitwich and is home to the Worcestershire Wildlife Trust's Christopher Cadbury Wetland Reserve. This complex of wetlands includes a series of salt-spring-fed flashes covering less than 10ha. Upton Warren's lakes are of recent origin having developed following subsidence during the second half of the 20th century. At that time other saltmarsh habitats were present in the general area such as the wetlands along the Droitwich Barge Canal, regrettably now much diminished. These wetlands and those that preceded them over the centuries may have provided continuity of saline habitat and so a source of species for colonisation.

As breeding birds, Avocets Recurvirostra avosetta became extinct in Britain in 1842. Their recolonisation of England in the early 1940s was a matter of good fortune as the defence of the realm required the flooding of vulnerable parts of the Suffolk coast. Amongst these natural defences the newly-made brackish lagoons provided an opportunity for wildlife. The subsequent establishment and rise in Avocet populations on the Suffolk coast is testimony to the efforts of conservationists and coastal landowners, not least the RSPB who adopted the bird as their emblem.

Once re-established as a British breeding bird, Avocets have spread across the southern and eastern coasts of Britain. In 2003 a solitary Avocet arrived at Upton Warren and was joined shortly after by a mate. That same spring four eggs were laid, all the eggs hatched and all the chicks fledged. Year on year the birds return and the breeding population grows.

In early summer the flashes at Upton Warren are a noisy place. Excited yells from the neighbouring adventure playground mix incongruously with the fluting calls of Avocets. Space for nesting and feeding is limited, with Canada Geese *Branta canadensis* and Black-headed Gulls *Chroicocephalus ridibundus* jostling for space but benefiting from a shared response to potential predators. The Canada Geese play an important role in helping to keep the vegetation short over the breeding season when cattle are excluded. Bare ground around the flashes not only provides habitats for nesting but also secures birdwatchers splendid views from the hides.

Avocets feed in the brackish flashes by sweeping the surface of shallows for invertebrates. The invertebrate communities of the salt wetlands are yet to be fully described; what is known indicates that whilst the flashes are only decades old they have already been colonised by specialists of saltmarsh habitats. Amongst the beetles recorded here is the scavenger *Enochrus bicolor* which is associated with the coastal marshes of the south and east of England. As a creature living on the edge of the marsh, *Enochrus* is capable of tolerating extreme variations in salinity. A study by Greenwood & Wood (2003) of the beetle's habitats on the Essex coast recorded the salinity of its pools over a four-year period. Salt concentrations ranged between 4.7ppt (parts per thousand) to 62.6ppt; seawater is about 35ppt. A pool filled on the highest tides and then evaporated down over summer may become hyper-saline, whilst the same pool over a rainy season may become almost fresh. To survive, there are advantages in being flexible to environmental change.

Saltmarsh plants can be found in the flashes but the site is not particularly species-rich. The short turf supports the community characterised by Lesser Sea-spurrey and Reflexed Saltmarsh-grass

Puccinellia distans. Radiating out from the saltmarsh are zones of vegetation reflecting decreasing degrees of salinity characterised by Grey Club-rushes, Celery-leaved Buttercup *Ranunculus sceleratus* and Spear-leaved Orache *Atriplex prostrata.* The most notable plant of the marsh is a moss, Heim's Pottia *Hennediella heimii.* This undemonstrative velvety mat is strongly associated with the landward edges of coastal marshes as it colonises disturbed

TOP: Celery-leaved Buttercup in the brackish marshes of Upton Warren.

ABOVE: The distinctive capsules of Heim's Pottia.

brackish soils. Cattle help to keep the habitat open for plants and birds alike. Without such pressure on the vegetation the likely outcome would be the dominance of reedbeds and wet woodland with the loss of the flashes' distinctive character.

Archaeological and geological studies emphasise that most species are naturally mobile given the right circumstances. If that were not so then recent geological history with its sequence of glaciations would have proved catastrophic. In 1961 Coope *et al.* published a report of their excavation of buried land surfaces at Upton Warren dating from some 40,000 years ago. The landscape at that time can be compared to tundra steppe, open grasslands that are warm enough in summer to provide lush vegetation for large animals but are cruelly cold in winter. The archaeological remains present a mammal fauna of Mammoth *Mammuthus primigenius*, Woolly Rhinoceros *Coelodonta antiquitatis*, Steppe Bison *Bison pricus*, Reindeer *Rangifer tarandus* and Norway Lemmings *Lemmus lemmus*. The flora contained many species regarded as typical of saltmarshes including Sea Plantain, Sea Arrowgrass, Sea-milkwort and Saltmarsh Flat-sedge *Blysmus rufus* as well as plants associated with alpine conditions such as Dwarf Willow *Salix herbacea* and Hoary Whitlowgrass *Draba incana*.

Caution is best applied in the interpretation of such evidence. What we regard as saltmarsh species may be present due to other environmental factors as illustrated by the presence of Slender Hare's-ear on the Malverns. The underlying geology helps the interpretation of the archaeological record as we know that it can, and does, support saline habitats. The chance survival of the evidence at Upton Warren, together with its fortuitous excavation, gives a snapshot of an historic landscape complete with wildlife. There is a treeless grassy plain with herds of large mammals gathering around saltlicks and water holes; such a scene invites parallels with the savannas of today's East Africa. Most of the plants and animals present in the tundra of Upton Warren would be recognisable to a modern naturalist; in ecological and evolutionary terms a mere forty millennia is no time at all.

Staffordshire

Saltmarshes in Staffordshire are not confined to the area around Pasturefields.

South of Burton-on-Trent is Branston. Maps of the 19th century illustrate a village near the River Trent with the advice 'liable to flood' engraved on the intervening fields. In December 1890 two local amateur naturalists, John Nowers and James Wells, read a paper to their local natural history society describing a saltmarsh at Branston including a preliminary species list. Their presentation also reported on water quality as the chemical composition of the local aquifers was a conspicuous part of the work of their associates. Messrs Bass and others were drawing on the mineral-rich waters beneath Burton to slake the thirsts of the growing cities of the Midlands. By the early 1880s Bass was the largest brewery in the world, exporting its pale ale throughout the Empire and, with much pride, into Édouard Manet's painting of *A Bar at the Folies-Bergère*.

Brackish Water-Crowfoot (left) has been lost from Branston Marsh whilst Saltmarsh Rush (right) has persisted.

Nowers and Wells produced an annotated map to accompany their paper which illustrated the exact ditches and pools within which they had found saltmarsh plants. Their account, published in 1892, suggests that groundwater extraction had already reduced the vigour of the springs, which had in turn resulted in a contraction of the saltmarsh. However, they were able to report the presence of Saltmarsh Rush, Wild Celery, Sea Club-rush and the Golden Dock *Rumex maritimus*. Despite its scientific name, the authors recognised this dock was not confined to saline habitats in its national distribution. Additional records from Branston by other botanists include Brookweed, Sea Aster and Sea-milkwort together with the Brackish Water-crowfoot *Ranunculus baudotii*; the cumulative species list suggests a saltmarsh of exceptional diversity.

In the early years of the 21st century the site of Branston's saltmarsh was studied by Mike Smith. Whilst still farmland, the detailed features of the landscape were found to be very different from those mapped by Nowers and Wells. A few of the ponds persisted but the pattern of ditches identified in 1890 had gone altogether. His investigations revealed that the late 19th-century landscape had been stripped for gravel, the flooded voids then drained and landfilled with pulverised ash; those ashfields were restored to farmland, initially growing arable crops, then sheep pastures and finally grazing for cattle. Ponds and ditches had been lost and new ponds had been created. With such a history of upheaval it seemed improbable that any of the original species could have survived.

The habitats known to Nowers and Wells have been overwritten but some of the species have proved remarkably resilient. Mike Smith located surviving populations of Saltmarsh Rush, Golden Dock and Grey Club-rush. Branston's floodplain pastures now form part of the green setting for neighbouring urban growth. There is a brackish character to the groundwater and the floodplain fluctuates over the course of a year. This water feeds seasonal ponds where cattle trample the margins and so promote niches for germination from the seedbank. The recent discovery of Brookweed suggests that conditions are increasingly favourable for saltmarsh species to reassert themselves.

Cheshire

In his novel of 1894, *The Queen of Love*, the Reverend Sabine Baring-Gould drew on the salt industry of Cheshire. Romance, high drama and moralising were played out as the circus arrived in the fictional town of Saltwich. Baring-Gould located his story on the River Weaver amongst the actual salt towns of Nantwich, Middlewich and Northwich. The impact of the salt trade on the lives and property of the people of the Wich towns had become something of a cause célèbre in Victorian society since the Great Subsidence of 6 December 1880. Thomas Ward, an eyewitness, recounted his experience to the Manchester Literary and Philosophical Society: 'so violent was the compression of air that it forced its way through every … contiguous district … showing itself in violent ebullitions in all the neighbouring pits, and where the earth was fractured causing a number of miniature mud geysers of 10 to 20ft in height. Much property was seriously damaged.'

Subsidence was an occupational hazard around the salt towns but the scale of the events of that December in Northwich was without precedent. A series of subterranean collapses captured the River Weaver and Wincham Brook before breaking surface to create the great lake of Witton Flash.

The geography of the lake was recorded by the Ordnance Survey who, unusually, annotated their map of 1899 with the precise date of when Witton Flash was surveyed; the clear inference being that, even after a decade, the land was still settling. The mapping of the new landscape recorded a complex series of saline habitats, with symbols

The Reverend Sabine Baring-Gould.

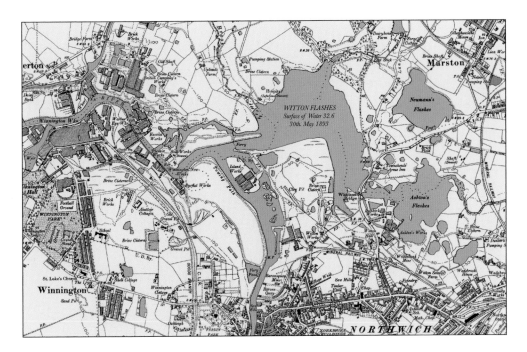

indicating marshes, swamps and rough grasslands. Unfortunately there appear to be no accounts as to what wildlife found a home there. We know from an account from Robert Holland of the Royal Agricultural College that in 1850 every crevice of the stonework around nearby Nantwich's saltworks was occupied with Lesser Sea-spurrey growing amongst both Common and Reflexed Saltmarsh-grasses. Whether the saltworks at Northwich shared a similar flora and what other species may have occupied the hollows of the Great Subsidence does not appear to have been investigated by contemporary botanists. Lord de Tabley's *Flora* of 1899 is remarkably silent on the botany of Cheshire's industrial heartland.

Witton Flash and the Great Subsidence. Adapted from Ordnance Survey map Cheshire XXXIV.NW, revised 1897, published 1899, reproduced by permission of the National Library of Scotland.

Contemporary sketches of the Great Subsidence.

The advent of industrialised saltworks not only generated prosperous towns but also, quite literally, undermined their very foundations. Baring-Gould contrasted the grave, high-minded civic leaders with the carefree residents of the poorer parts of town. In fictional Saltwich the old town hall had recently collapsed having been undermined and

Two streets had gone to rack and ruin, rifts had formed in the house-walls, and stacks of chimneys had fallen. The grave regarded this as a judgement on the gay who had inhabited them whilst themselves pumping and steaming and sending away and converting into gold the salt rock that underlay these habitations.

Having forced the circus onto waste ground near a flash, the performance was interrupted:

The central tent pole was seen ... to be sinking into the earth ... it appeared as though the pole were collapsing telescopically ... the crimson drapery was parted as though rent asunder by giant hands ... the terror was general. ... The circus and the ground about it were going down into an unfathomed abyss.

Whilst the vocabulary and biblical allusions are archaic, the melodrama would not be out of place in a contemporary exposé of the iniquities of unregulated capitalism. At the heart of the issue was a technique called wild brine pumping. The salt industry had developed ways of forcing steam and hot water into the rocks then pumping up the resulting brine. At the same time other cavities were being created through conventional mining. With good fortune the dry mines were effectively sealed from the brine streams. This separation could not be guaranteed and the net result of their meeting was catastrophic.

Subsidence due to water flowing through rock salt is a natural phenomenon and occurred long before the development of the salt industry. The geological processes are similar to those of chalk and limestone, which similarly create karstic features. Rostherne Mere near Knutsford is the largest of the Cheshire meres. It is a deep, steep-sided lake that formed towards the end of the last ice age through subsidence following glacial meltwater dissolving rock salt. The mere is now a National Nature Reserve of principal interest for its wintering waterfowl. If the lake has any connection with the underlying rocks this is not detectable through the species that live there.

Naturally-dissolving salt deposits have resulted in subsidence throughout recent centuries. A graphic account of such an event to the west of Nantwich was made by William Jackson in a report to the Royal Society of 1669:

> That near a place of My Lord Cholmondley's called Bilkely, …
> without any earthquake, fell in a piece of ground about 30 yards over,
> which a huge noise, and great Oakes growing on it fell with it together;
> which hung first with parts of their heads out, afterwards suddenly
> sunk down into the grounds so as to become invisible. Out of the pit
> they drew brine with a pitcher tyed to a cart rope, but could then find no
> bottom with the ropes they had there.

Much has been done to stabilise and infill the results of the Great Subsidence. The largest salt lake surviving from this period is Neumann's Flash. Highly alkaline waste from a soda ash works has been used to reduce its depth, leaving an expanse of water characterised by the extremes of salt and soda, the chlorides of sodium and calcium. Today Neumann's toothpaste-smooth margins and alkaline crust are reminiscent of the salt lakes of continental deserts.

Neumann's Flash.

The wildlife of Neumann's Flash reflects its history and management. Extensive growths of reed and wet woodland have developed on the unstable sediments. Beds of Purple Glasswort *Salicornia ramosissima* colonised the shore as part of the natural restoration of the wasteland. Glassworts are a complex suite of species of succulent plants occupying the early-stage succession habitats of saltmarshes and coastal mudflats. The slender, salty, brittle stems are regarded by some as a delicacy and eaten as samphire. The population of glassworts at Neumann's was without parallel in England's inland saltmarshes but is one of the component features of equivalent places in France and Germany. At Neumann's Flash succession has overtaken the Glassworts but other saltmarsh species persist such as Sea Aster, Horned Pondweed *Zannichellia palustris* and Reflexed Saltmarsh-grass. This is a large lake, and given the perils of botanising over highly unstable ground there remains the possibility that additional saltmarsh species may yet be discovered.

Neumann's Flash and Northwich are at the heart of salt country. In recent years the Anderton Nature Park has been established to help the restoration of this post-industrial landscape. The park provides a positive use for land unsuitable for redevelopment. Amongst the amenity grasslands and tree-planting schemes there are salt springs breaking out as saline groundwater reasserts its influence. The springs are not spectacular but are sufficient to support familiar plants from other inland saltmarshes to testify to the land's geological and industrial heritage. Most remarkable is a population of Stiff Saltmarsh-grass *Puccinellia rupestris*, a national rarity associated with scuffed habitats in the drier saltmarshes of southern England.

A few kilometres to the south of Northwich the salt mine at Winsford provides much of the rock salt used to de-ice Britain's roads. The engineering there is of heroic proportions as giant machines and conveyors mine to a depth of some 140m, creating over 200km of tunnels. Great mounds of salt are piled around the mineshafts, the insoluble elements conveniently forming a waterproof thatch to prevent the entire stock dissolving. On the opposite bank of the River Weaver the natural topography gives rise to reliable salt springs. The bluff is impressively steep, almost an inland cliff with a spring-line curiously aligned on its upper slope. Some of the springs are freshwater and are marked by a confusion of marsh-orchids *Dactylorhiza* spp. The salt springs are characterised by dark metallic stains and crusts of crystals, their numerous seepages gathering to feed a rivulet set within

a briny duct. Sea-milkwort and Brookweed grow along its edge as it flows down the slope into a pool, which in turn feeds a small channel. Grey Club-rush lines this salt-stream, which has all the characteristics of a minor creek winding across a coastal grazing marsh. Other seeps at the foot of the slope are encrusted by Stoneworts *Charophyta* and punctuated by squat plants of Sea Aster and the familiar combination of Lesser Sea-spurrey and Reflexed saltmarsh-grass. The esoteric delights of these saltmarshes are accompanied not by the sounds of the sea but by the clarion call of klaxons.

The minehead at Winsford from across the River Weaver.

Life on the verge

Winsford's rock salt provides a postscript to this exploration of inland saltmarshes. By reading historical floras one can gain an insight into the character and distribution of former habitats. Species accounts made by previous generations of naturalists can be collated to describe naturally landlocked saltmarshes, a British habitat which is mostly lost and almost forgotten. The provisional inventory of such marshes suggests that what survives today is an impoverished relic of a once widespread wetland. Such an analysis cannot be made of more modern floras.

The widespread practice of spreading salt onto the road network dates back to the 1970s. At first the expansion of saltmarsh species into

A liberal sprinkling of salt.

Common Scurvygrass in the marshes below Cader Idris, Gwynedd.

brackish verges was regarded as a curiosity. However the diversity and dynamism of the changes it stimulated soon established a novel line of botanical inquiry.

Danish Scurvygrass *Cochlearia danica* is the most widely distributed saltmarsh plant along roadsides. It is unhelpfully named as it is neither a grass nor has it any particular affiliation with Denmark. Scurvygrasses are small, annual, acrid-tasting members of the cabbage family. Danish Scurvygrass has proved to be the most adventitious of the many *Cochlearia* species that occur naturally in saltmarshes. In less than forty years it has colonised most of the trunk roads south of Glasgow through the advantageous life cycle of germinating in winter and setting seed before mowing. All Scurvygrasses share the antiscorbutic quality of being rich in vitamin C and so provide protection from scurvy. Historically they were important ingredients in treating dietary deficiencies. Largest of them all is Common Scurvygrass *Cochlearia officinalis*, which is an occasional occupant of salted verges. Common Scurvygrass was cultivated for its medicinal properties up until the mid-19th century. The importation of cheaper and more palatable

lime juice put an end to this home-grown cure. Common Scurvygrass's specific name *officinalis* is derived from *opificina*, an adjective originally relating to workshops and then later used to describe stores of herbs in monasteries and apothecaries' shops before being adopted by botanists to describe any plant used in medicine.

Two other widespread colonists from the 1970s are Lesser Sea-spurrey and Reflexed Saltmarsh-grass, both of which are components of a saltmarsh community naturally occurring in the upper reaches of a marsh. This association occupies open ground which may be a tidal pan or poached ground around a cattle path. Where turf is cut for the horticultural trade, such as on the Solway marshes, this is the community that colonises the exposed soil and starts the process of re-establishing a fine grassy sward. In road verges the two usually grow in partnership, the Sea-spurrey running to seed as the Saltmarsh-grass comes into bloom.

As Lesser Sea-spurrey and Reflexed saltmarsh-grass have spread inland so have other species from a wide range of saltmarsh types. The progress of colonisation can be traced through the letters and contributions to *BSBI News*. Sea-milkwort grows amongst Lesser Sea-spurrey and Reflexed Saltmarsh-grass on the A93 beneath Ben Gulabin in the southern foothills of the Cairngorms. Curved Sedge *Carex maritima*, a rarity of the far north, has appeared in the salty verges of Lewis and Harris in the Outer Hebrides. In central and southern

Shrubby Sea-blite at home on the North Norfolk Coast.

England colonies of Shrubby Sea-blite *Suaeda vera* have colonised the M6 and A13; this is a species associated with the Mediterranean-influenced gravelly marshes of the East Anglian and southern coasts. The expectation of fresh discoveries has stimulated botanists into acts of daring-do with reckless accounts of forays along trunk roads. What was once frustrating congestion is now an opportunity to inspect, if not access, the sanctums of central reservations and hard shoulders.

The means by which saltmarsh plants colonised verges was the subject of Nick Scott's doctoral thesis in the early 1980s. Through novel methods, including setting roadside seed-traps and filtering detritus from a carwash, he established that the pattern of colonisation mostly followed the direction of the traffic. Scott's seminal work was partially published in the academic press but only gained its rightful prominence some twenty years later when in 2003 he contributed to the debate in *BSBI News*.

Roadside saltmarshes are not always the result of salting roads. In 1994 a rich saline community was discovered by the Hartside Cafe, high on the Pennine moors between Penrith and Alston; the flora including Sea Arrowgrass, Sea Milkwort, Distant Sedge, Saltmarsh Rush and Sea Plantain. Local enquiries established that the worn surrounds of the lorry park had been landscaped with turf from a Solway saltmarsh. The ecological stresses of growing at an altitude of 580m combined with sheep grazing and the workings of a transport cafe had successfully sustained the marsh plants for over twenty years. Given the scale of the horticultural trade in turves from saltmarshes, it is surprising that such sites are not more common.

The wild flowers of salty verges are cosmopolitan, with occasional glamorous additions from abroad. A conspicuous newcomer is the Summer-cypress *Bassia scoparia* which originates in the salt steppes of central Asia. This close relative to our native goosefoots *Chenopodium* spp. and oraches *Atriplex* spp. has long been cultivated as an eye-catcher in garden bedding schemes. As a fast-growing annual the Summer-cypress can develop into a thigh-high elongated dome of thread-like leaves. Even when glimpsed from a moving car the shape is unmistakably like the bearskin cap of a Grenadier Guard, or alternatively Cousin Itt from the Addams family. As summer progresses the foliage blazes red and purple, further ensuring it is unlikely to be overlooked. Through the opening years of the 21st century Summer-cypress spread along the motorways of Hampshire and Kent as far north as Yorkshire and westward

Summer-cypress in a garden setting.

to Somerset. Colonisation was no doubt assisted by the dead but seed-bearing remains being bowled along as tumbleweed. In recent years the original vigour appears to have waned. There is even a suggestion that Summer-cypress is being outcompeted by the far less flamboyant, but native, Grass-leaved Orache *Atriplex littoralis*.

Through digging and spreading rock salt we have redistributed enough of Britain's solid geology to diversify the cast of the seasons. It is the blushed blooms of Danish Scurvygrass that offer the car-bound commuter a promise of spring.

Danish Scurvy-grass in the Peak District of landlocked Derbyshire.

On the wind

S alt is a mobile compound in the form of windblown crystals and in solution as spume and spray. Anyone who has walked an exposed coastal path on a blustery day will have experienced the accumulation of fine films of salt on their skin and spectacles. When coastal vegetation is consistently exposed to salt-winds then perched saltmarshes will develop well above the limits of the tide.

On the morning of 14 January 1803 Mr R A Salisbury Esq instructed his servant to remove a hoar frost from the recently cleaned windows of his house. He recalled in his presentation to the Linnean Society (1805):

> *When the servant who was sent to remove it, came and told me it was salt, I was astonished and even more so upon going out to find this substance almost as abundantly deposited in the garden and neighbouring fields. I was totally at a loss to account for so curious and unusual a Phenomena, and the next morning with my gun in my hand I walked over a circuit of twenty miles, before I ventured to trust the evidence of two of my senses, and bring some branches from the hedges still salt to Sir Joseph Banks.*

Robert Salisbury lived in Mill Hill, Middlesex, in what is now the London Borough of Barnet. His home was on high ground some 15km from the tidal Thames and nearly 60km from the open sea. That January a series of prolonged easterly gales drove salt far into the home counties of southern England. The impacts were short-lived; Salisbury observed the effects of scorch from frost and salt diminishing with the seasons as a mild climate and plentiful rain restored his garden to its former state.

OPPOSITE PAGE:
Spume, Compton Bay,
Isle of Wight.

With increased latitude and exposure the effects of wind-borne salt are longer lasting. Hermaness is the most northerly headland of Unst, one of the Shetland Islands, and is host to the northernmost intertidal marsh in Britain. Plant communities of the intertidal spread out from above the tideline to follow the adjoining cliff right up to its crest at 170m. The island's rocky headlands are subject to frequent storms driving spume and spray across the cliff-tops. In this hyper-oceanic climate with mild stormy winters and cool windy summers the air is consistently humid and the soils invariably wet. Cliff-top vegetation reflects the competing influence of brine with freshwater. Closer to the cliff edge the effects of sea salt are permanently expressed, the enriching effect of the minerals supporting a zone of nutritious Red Fescue *Festuca rubra* before freshwater influences replace it with acidic grasslands and blanket bog.

Such conditions challenge artificial classifications of habitat as maritime cliff communities merge into saltmarshes. Saltmarsh Rush *Juncus gerardii* and Sea Arrowgrass *Triglochin maritima* are the most consistent saltmarsh plants to be found in perched marshes on northern cliffs. The recently completed Scottish Saltmarsh Survey by Tom Haynes and colleagues (2016) recommends using these species to separate perched saltmarshes from other maritime cliff communities.

In such extreme environments the divisions between cliff and marsh are subtle and artificial. Elsewhere on the mainland coastal cliffs support species that would be regarded as characteristic of saltmarshes were they found in a different topography; abundant Sea Plantain *Plantago maritima* and Thrift *Armeria maritima* grow in a fine grass sward alongside Common Scurvygrass *Cochlearia officinalis* and the Small-fruited Yellow-sedge *Carex oederi*. It is in this community that the Scottish Primrose *Primula scotica* occasionally occurs. Derek Ratcliffe, in his 1977 review of British coastal plants, regarded the Scottish Primrose as being a species of both cliff and saltmarsh. His reasoning is sound even if discomforting to conventional wisdom.

It is not necessary to be exposed to the full force of the Atlantic for perched saltmarshes to develop. Along the relatively low coast of Angus a highly diverse saltmarsh has formed above the tidelines at Usan near Boddin Point where sea spray is still sufficient to support briney conditions. Ingram & Noltie (1981) recorded the flora here as having both saline and brackish elements supporting distinctly northern species including Saltmarsh Flat-sedge *Blysmus rufus* and the Curved Sedge *Carex maritima*.

Scottish Primrose.

BELOW: Perched saltmarsh on the cliffs at Usan.

Much further south in the shelter of the Dee Estuary a perched saltmarsh has formed on Hilbre Island. Hilbre is the largest of the rocky outcrops lying between the Welsh and English coasts accessible on foot at low tide. A hollow in the island's rock platform contains a flood pasture community of Silverweed *Potentilla anserina* and Marsh Pennywort *Hydrocotyle vulgaris* which grades down through Saltmarsh Rush *Juncus gerardii*, Brookweed *Samolus valerandi* and Sea-milkwort *Glaux maritima* then finally into Sea Club-rush *Bolboschoenus maritimus* with Spiked Water-milfoil *Myriophyllum spicatum*. The whole marsh is just a few metres across, elevated some 10m above the estuarine flats.

Whilst wind and weather favour perched saltmarshes in the north and west of Britain, these habitats are not confined to hard rock coasts. The cliffs of southern England are also exposed to salt spray and rainfall.

Hilbre with its marsh flooded by winter storms.

Sea-heath blooms above
the Needles.

Chalk is hard enough to form sheer cliffs but not as resistant to erosion as many other sedimentary rocks. Such cliffs are perilous to the inquisitive naturalist, friable enough to crumble and high enough for a fatal fall. Fortunately the perched saltmarsh that crowns the Needles Headland on the Isle of Wight can be approached in complete safety. The fortifications constructed in the late 19th century include a searchlight post accessed by a brick-lined tunnel from the Battery's parade ground. From within the cage at the tunnel's end one can appreciate stands of Thrift and Sea-purslane *Atriplex portulacoides* growing with Sea-heath *Frankenia laevis* and Annual Sea-blite *Suaeda maritima* in a community reminiscent of the saltmarshes of Rye Harbour in Sussex. Here at 60m Sea-heath is found at the western limits of its natural distribution in Britain as well as its highest elevation.

From the Needles one can look westwards to Dorset's Jurassic Coast where short-lived wetlands form in the seepage-fed hollows of landslips. The ephemeral nature of such habitats tends to favour relatively mobile

TOP: Slender Centaury has its British stronghold on the brackish seepages at Eype, Dorset.

ABOVE: The Long-horned General.

organisms rather than the development of mature saltmarsh communities. Dorset's brackish cliff pools support species associated with coastal grazing marshes such as Sea Clover *Trifolium squamosum* and the soldierfly, the Long-horned General *Stratiomys longicornis*. The continuity of open ground produced by brackish mudslides sustains conditions for the Slender Centaury *Centaurium tenuiflorum*. This pale-flowered member of the gentian family has always been a rarity in Britain with a handful of historical locations in the saltmarshes and coastal grasslands of the south coast. The restricted range in Britain is probably due to climate, the continental distribution of the species being saline habitats throughout the Mediterranean into the Crimea and Iraq.

In contrast the Shore Dock *Rumex rupestris* has its global stronghold on the Welsh and West Country coast. Presented by Lousley and Kent (1981) as one of the world's rarest docks, its preferred habitats are varied

Sheltered coves on the South Hams, Devon, support marsh communities with Shore Dock.

and include an extreme and exceptional form of saltmarsh. The marshes associated with Shore Dock tend to be just a few square metres of partially vegetated turf perched on the highest parts of a storm beach. What little soil accumulates is periodically soaked in salt whilst being flushed by freshwater runnels. The community supports typical saltmarsh plants such as Sea Beet *Beta vulgaris* ssp *maritima*, Brookweed and False Fox-sedge *Carex otrubae* growing in a matrix of fine grasses. Some of the richer communities, such as on the South Devon coast, include Long-bracted Sedge *Carex extensa* and rock sea-lavenders *Limonium* spp. By their very nature the habitats of Shore Dock are unstable. If storms do not rearrange the beach then the marsh may become overgrown by coarse grasses and reeds. As a poor competitor the Dock is at risk of being crowded out in the absence of grazing animals keeping the habitat open. If favourable conditions persist then populations of the Dock can grow into hundreds of plants as the deep-rooted perennial produces an abundance of seeds, a successful strategy for persisting in such unstable places.

In the Highlands

S altmarshes are a common component of the coastal habitats of the Scottish Highlands. These marshes are highly diverse and support species found nowhere else in Britain. The presence of saltmarshes within the extensive habitat complexes of the Highlands is of interest in its own right as well as enhancing our understanding of the fragmented saltmarshes of lowland Britain.

A northern marsh

Eddrachillis Bay, to the north of Assynt, is a coastline of low cliffs and uninhabited islands, sparsely peopled around its bays and sheltered sea lochs. Loch Nedd and its eponymous township lie on the southern shore in the lee of rocky headlands amongst the foothills of the Quinag. The western cliffs of that mountain feed the headwaters of the Abhainn Gleann Leireag, providing a constant flow of freshwater into the loch. A simple saltmarsh has formed here where the river meets the coast; there are a few tens of square metres of intertidal turf, an unremarkable minor component of the hundreds of square kilometres of the open country of Assynt.

There is no such thing as a typical loch head and so Loch Nedd is an arbitrary choice to demonstrate saltmarshes in these latitudes. Chance has it that Nedd does not have a well-developed turf of *Fucus muscoides*, a seaweed that adopts a terrestrial habit on the neighbouring marsh at Oldany. Similarly, Nedd is not particularly rich in liverworts and mosses which are an important feature in some other Highland marshes. What Nedd does exhibit are common features of northern saltmarshes, being mostly rocky with gin-clear waters as there are few of the fine suspended sediments that feed the marshes of more

OPPOSITE PAGE:
Kentra Bay, Argyll.

southern climes. Pioneer marsh vegetation including glassworts *Salicornia* spp. are absent, the foreshore and creeks being populated instead by seaweeds. The absence of these pioneer communities is due to substrate, not climate; glassworts will grow in suitable habitats and can be found further north in muddier marshes. In Cameron's 1883 dictionary of Scottish Gaelic the glassworts were not assigned a name of their own but were lumped along with other closely related species into the category of 'Praiseach na Màra', pot-herbs of the seaside. Such a treatment suggests that they were neither common nor valuable enough to require more specific treatment.

The Gaelic language embraces most of the species at Nedd. Some plants, particularly those with utility or cultural associations, have names reflecting a history of commonplace usage. 'Bilearach' is Gaelic for eelgrass *Zostera* spp., which is recorded by Evans *et al.* (2002) from Oldany's shores. This long-standing name reflects the value placed on the grass for thatching and stuffing mattresses.

Loch Nedd's saltmarsh has formed between the rocks of the foreshore and over cobbles at the mouth of the river. In a few paces you can cross from wooded crags, through the marsh and into the seaweed-lined creeks. The first impression is a sward of fine-leaved grasses with abundant 'Fèisd Ruadh' Red Fescue *Festuca rubra* and 'Feur Rèisg Ghoirt' Common Saltmarsh-grass *Puccinellia maritima*.

The head of Loch Nedd.

The sedges and rushes are equally delicate, the northern character being expressed through the abundance of 'Seisg nam Measan Beaga' the Small-fruited Yellow-Sedge *Carex oederi* together with 'Seisg Rèisg Ghoirt' Saltmarsh Flat-sedge *Blysmus rufus*.

The herbs of the intertidal are those common to most British saltmarshes, with 'Lus na Saillteachd' Sea-milkwort *Glaux maritima* and 'Slàn-lus na Mara' Sea Plantain *Plantago maritima* together with 'Neòinean Cladaich' Thrift *Armeria maritima*. The Gaelic for Thrift has been variously translated as the beach wave or the beach daisy; in summer its flowers define the higher intertidal zone with a picotee of pink.

In amongst the rocks of the marsh grows a near-prostrate form of Scurvygrass *Cochlearia* sp. Determining Scurvygrasses to species level is seldom simple, a task which is magnified where there are environmentally induced variations. The material at Nedd could be Atlantic Scurvygrass *Cochlearia atlantica* or alternatively 'Carran Albannach' Scottish Scurvygrass *Cochlearia officinalis* ssp. *scotica*. The trouble with these closely related associates of 'Am Maraiche' Common Scurvygrass *Cochlearia officinalis* is that botanical descriptions are hedged about with qualified terminology – slightly, normally, frequently, usually. Such an accumulation of relative judgements offers little help to a simple mind seeking features to determine a diagnosis. Advice by Tim Rich (1991) is that *Cochlearia atlantica* may be

LEFT: 'Seisg nam Measan Beaga' Small-fruited Yellow-Sedge.

RIGHT: 'Seisg Rèisg Ghoirt' Saltmarsh Flat-sedge.

regarded as a species with a world distribution confined to the British Isles. Scottish Scurvygrass was regarded as a species by Druce who named it *Cochlearia scotica*, a decision subsequently overturned by more critically minded authorities. As fresh thought and further evidence become available then things may become clearer; alternatively we may just accept that the concept of species is artificial and the need to name names is as much cultural as scientific. The Scurvygrasses are interesting for their plasticity even if we cannot assign every plant a label in our desire for order.

The character of Nedd's marsh changes not only with elevation but also with the degree by which it is infused with freshwater. It is worth searching flushed turf for 'Falcair Mìn' Chaffweed *Centunculus minima*, a member of the primrose family reputedly bearing the smallest flower of any terrestrial plant in Britain. In contrast, the largest component of the saltmarsh at Nedd is also found in this zone; this is 'Sèimhean Dubh' the Black Bog-rush *Scheonus nigricans*. Translations of the Gaelic vary in reflecting the shiny qualities of the Bog-rush's flowerhead or alternatively its use as straw. Less convincing is an antiquarian account published by Cameron (1883) of fairy folklore with Bog-rush stems providing shafts for elven arrows, tipped with white flint and bathed in hemlock.

An intertidal specimen of 'Sèimhean Dubh' Black Bog-rush.

The landward end of Nedd's marsh sequence concludes with brackish grassland speckled with nomenclature-defying 'Lus nan Leac' eyebrights *Euphrasia* spp. and 'Beàrnan Brìde' dandelions *Taraxacum* spp. An uncommitted botanist can opt to appreciate the diversity of such complex species without feeling obliged to allocate names in whatever language may be most fitting.

Neither colloquial nor scientific names are necessarily reliable indicators of the true nature of a plant. Thrift and Sea Plantain share the scientific specific of *maritima* yet in Assynt their distribution is not confined to the maritime. When entering the parish from the south, the road takes you under the limestone cliffs of Creag Sròn Chrùbaidh. On these crags luxuriant clumps of Sea Plantain abound amongst Mountain Avens *Dryas octapetala* and the Dark-red Helleborine *Epipactus atrorubens*. In contrast Sea Plantain is reduced to little more than a vegetal crust in the quartzite fell fields of Glas Bheinn, the neighbouring peak to the Quinag. Here the Plantain grows alongside tight mounds of Thrift in amongst Trailing Azalea *Kalmia procumbens* and Alpine Clubmoss *Diphasiastrum alpinum*. Environmental stresses at these latitudes and altitudes extend the distribution of saltmarsh species far beyond their *maritima* habitats into affiliations with the Arctic and the Alpine.

'Slàn-lus na Mara' Sea Plantain grows on the limestone cliffs of Creag Sròn Chrùbaidh.

The most demanding of Arctic-Alpine saltmarsh species is not known from Loch Nedd but can be found in a handful of sites along the shores of the Highlands and Outer Hebrides. The local name for this plant is 'Seisg Bheag Dhubh-cheannach', the Gaelic reflecting the conspicuous small dark clustered flowerheads. I first saw Curved Sedge *Carex maritima* whilst hiking beneath Monte Rosa in the Swiss Alps where it introduced an air of sobriety into the exuberance of the alpine summer. Amongst the skeletal soils of glacial outwash there appeared to be a scatter of freeze-dried blackberries thoughtlessly discarded by some picnicker. On closer investigation those dark clusters were flowerheads, disproportionately chunky compared to the modest tufts of curving leaves, each isolated sprig arising from a net of rhizomes.

From a European perspective the Curved Sedge populations of the Alps are outliers of a primary distribution along the coasts of the Arctic and Scandinavia. The Scottish populations may be seen as a southern expression of this Arctic influence, possible survivors from when glaciers last moved north. There are a number of locations for Curved Sedge along the coast of the Pentland Firth including a sheltered bay below Bettyhill. The abundance of this population is revealed as the

Curved Sedge habitat below Monte Rosa.

falling tide exposes the sandflats; the marsh is peppered with knots of flower anchored against the current. Even where growing at its greatest density the sedge does not cover the shore; bare sand remains a dominant element in the community. In amongst the Curved Sedge grow saltmarsh familiars, Thrift, Sea Plantain, Sea-milkwort and Small-fruited Yellow-sedge, but nowhere do these additions combine to become a persuasive sward; they are spartan fellow travellers at an extreme limit of Britain's saltmarsh vegetation.

Before moving on from the northern shore of mainland Scotland it is worth considering just one more sedge, *Carex salina*, the newest recruit to Britain's saltmarsh flora. In 2004 Keith Hutcheon was surveying a loch-head marsh at Morvich on Loch Duich beneath the Sisters of Kintail. Here he found a sedge, a nameless grassy sedge in a grassy marsh. Any botanist will admire Hutcheon's persistence with the question 'what is it?'; in my experience such stuff tends to

'Seisg Bheag Dhubh-cheannach' Curved Sedge.

Curved Sedge habitat at Bettyhill.

Saltmarsh Sedge.

Morvich marsh lies below a
burial ground of the Clan McKay.

get composted. The specimen was eventually determined as *Carex salina*, the Saltmarsh Sedge; Hutcheon enjoys the rare distinction of finding a wild plant new to Britain.

Saltmarsh Sedge is a species known from Norway and north-western Russia together with Canadian populations from Hudson Bay to the east coast. The discovery of a colony in Scotland was therefore intriguing but not improbable. To add interest *Carex salina* is a species which has emerged from the stabilisation of hybrids of *Carex paleacea* and *Carex subspathacea*, neither of which has ever been recorded from Britain. The species may have emerged elsewhere and found its way to the Highlands; alternatively its parents may have been present in Scotland but are either extinct, or persist undetected. As the Morvich marsh had been surveyed in the early 1970s there was speculation that the sedge may have

colonised very recently, a flattering but implausible suggestion; if only every survey really could record every species.

In 2006 likely-looking sites along the west coast were surveyed for Saltmarsh Sedge and later the media promoted an appeal for citizen science to find more populations, all to no avail. One of the most effective ways of prompting additional discoveries is to go to press stating that a species is confined to a single site, and such a suggestion was published by Dean *et al.* in 2008.

Between 2010 and 2012 Scottish Natural Heritage commissioned a systematic survey of Scotland's saltmarshes which was produced as a report by Tom Haynes in 2016. As part of that survey the dedicated fieldwork of Ian Strachan found the Saltmarsh Sedge at Loch Nevis, a remote marsh in the roadless country of Knoydart. Another population turned up at Torrisdale Bay not far from the university field station at Bettyhill. Two populations were also located at the head of Loch Sunart.

Within a decade of its first discovery the Saltmarsh Sedge has gone from being an extreme rarity to a species with a convincingly natural distribution across the northern and western shores of the Highlands. It is reasonable to expect additional sites to be found as

Saltmarsh at the head of Loch Sunart.

botanists become familiar with the plant and its habitats. The evidence now suggests the Saltmarsh Sedge is a long-standing but overlooked member of our native flora.

I had driven past the Sunart marshes on many occasions, never stopping on my way to the riches of Morvern. On hearing of its discovery I paid my respects and felt rather underwhelmed for my efforts. The sedge is a smaller, more shyly flowering version of the closely related Estuarine Sedge *Carex recta*, another of Scotland's rarities with links to higher latitudes. Estuarine Sedge is robust enough to form dominant sedge-beds, visible at Bonar Bridge from the car as you cross the Kyle of Sutherland; its smaller cousin needs searching for. Sedges may be an acquired taste but the pursuit of them beguiles you into lingering in wonderful places.

Saltmarsh woods

Loch Nedd is a wooded sea loch where the trees are rooted above the tides. This is the usual circumstance where woodlands meet saltmarshes, as tree roots and saltwater seldom mix. In temperate regions there is no equivalent to the mangroves of the tropics but there are exceptional circumstances where woodlands grow into the intertidal zone.

Britain's largest estuarine alder wood can be found at The Mound in the headwaters of Loch Fleet, north of Dornoch. The Mound alder woods formed upstream of a causeway built in the early 19th century. Thomas Telford designed the causeway, a rather crude piece of engineering compared to his prestigious projects along the Caledonian Canal and across the Menai Straits. The Mound is a simple kilometre-long bank of rock with tide gates at one end.

The causeway was completed in 1816 on the instructions of Lord Stafford; he was a successful industrialist who had recently married Duchess Elizabeth, the heir to the Sutherland Estates. Having married into property he set out on a process of aggrandisement and improvement, purchasing additional land to create an estate of some 485,000ha (*c.*1.2 million acres) taking up nearly all of the county of Sutherland.

The Mound functioned reasonably well as a road but it modified the estuary of Loch Fleet. The tidal reaches of the River Fleet and the Abhainn an t-Stratha Charnaig were now restricted by the tide gates. At low tide the waters of these substantial rivers needed releasing into

the sea loch and at high tide the gates could be closed to limit saline incursions. The engineering proved to be imperfect and so the tide was disrupted but not excluded; saltwater continued to flow upstream but without its previous vigour and scour.

Early 19th-century maps show saltmarshes forming upstream of the tide gates as trapped silt overwhelmed the tidal flats. Later maps illustrate progressive colonisation by trees. A study of the tidal alder woods by Hendry & Edwards (2012) has demonstrated woodland expanding across the marsh in discrete pulses from the 1830s onward until the mid-20th century. In all there are about 180ha of alder woods experiencing a degree of salinity. Hendry and Edwards found that within some of the younger cohorts of Alders *Alnus glutinosa* there are groups of far older trees. It is possible that, before engineering, the estuary of the River Fleet had similarities with the tidal Conon where it flows into the Cromarty Firth by Dingwall. The braids of the Conon estuary flow between wooded islands of Alder, willow *Salix* spp. and oak *Quercus* spp. Had such stands of Alder been already present on the Fleet then Telford's engineering would have inadvertently changed conditions in their favour. Today saline conditions at The Mound are in a constant state of flux depending on the weather and the operation of the tide gates. The Hendry

Sluices regulate the passage of the tide through The Mound.

77

The estuarine alder woods
of Loch Fleet.

and Edwards study was commissioned in response to concerns that parts of the wood were dying. The study confirmed this was so; trees stressed by variations in salinity were suffering from an infection of the fungus *Phytophthora alni*. This die-back however was not affecting the whole wood; there are still vigorous groves of young trees. In growing on the cusp of tolerance the alder woods of The Mound are sensitive to the slightest of environmental changes.

The alder woods of the Fleet and Conon are both associated with the estuaries of major rivers. Not all intertidal woods develop in such sites.

At the opposite end of the Highlands to Loch Fleet is Knapdale where Loch Sween flows into the Sound of Jura from the Kintyre peninsula. The tidal regime of this part of the Inner Hebrides is complex, with the labyrinth of islands, sea lochs and peninsulas generating one of the lowest tidal ranges in Britain.

Loch Sween fills the low-lying ground between ridges of grit stones and basaltic rocks. The main loch and its many offshoots run along parallel alignments through wooded hills. Set apart from the perilous Sound of Jura, here is a place of quiet waters where the topography of the valleys is mostly sheer, with a few gentler slopes enabling saltmarsh to develop in a narrow band along the shore.

OPPOSITE PAGE:
Loch Sween.

Iris communities are tolerant of being occasionally flooded by seawater.

From a foothold on the marsh the view along the loch is of four abrupt divisions. The lowest shore has no plants; rock and shell provide holdfasts for seaweeds exposed at low tide. There is then a step up from the cobbles to the familiar combinations of fine grasses and herbs of a Highland saltmarsh. In its turn this short turf yields to tall herbs, Yellow Iris *Iris pseudacorus*, Meadowsweet *Filipendula ulmaria* and Hemlock Water-dropwort *Oenanthe crocata*, so in season the foreshore seam of flowering Thrift is backed by herbaceous creams and gold. The stands of Iris and their allies comprise a fen community which is well within the intertidal zone. Shadowing them all is Alder in a sopping freshwater swamp. Amongst its branches is a stomach-high strand of tidal debris, remnants of the last time seawater swept through. The survival of intertidal woods depends on them being constantly doused with freshwater whilst being tolerant of an occasional sousing of salt.

Coastal mires

Visitors to the northern hills of Knapdale are directed to a bluff where there are spectacular views over Mòine Mòhr, the Great Moss of the River Add. The Add flows through the raised bogs of the Mòhr, meandering through tidal reaches into saltmarshes by Crinan Ferry. Over the last five millennia the Mòhr has grown into an ancient tidal basin with a scatter of islands, some landlocked within the peat, others straddling today's coastal sediments and saltmarshes.

OPPOSITE PAGE TOP:
Crinan Ferry.

OPPOSITE PAGE BOTTOM:
Saltmarsh and raised bog merge at Mòine Mòhr.

In one sweeping panorama you can see transitions of the bog into woods, farmland and saltmarsh, a rare example of a complete and complex estuarine landscape.

The presence of peatlands of various types along the Highland coast is not particularly unusual. In many coastal wetlands, saltmarshes grade into peaty soils, so generating unlikely neighbours such as Saltmarsh Flat-sedge with cottongrasses *Eriophorum* spp., or Yellow-sedge with Heather *Calluna vulgaris*. Mòine Mòhr differs in scale with an estuarine frontage to the bog of several kilometres. It is the most complete transition of peatland and intertidal surviving in Britain.

The general pattern of marsh development over recent millennia is of Mòine Mòhr growing vertically, adding a millimetre or so in elevation to the peat dome each year. Saltmarshes progress in a different direction, the seaward migration of marsh compensating for the gradual elevation of the land. Local erosion and deposition continually nuance these strategic trends. Saltmarsh communities have established on peat slurries, settling in quiet bays along the shore; elsewhere the intertidal is crossed by seepages supporting bands of bog mosses to the limits of their saline tolerances. The Add severs peat from its meandering banks, some blocks as big as freezers, and floats them downstream to founder on the intertidal.

Mòine Mòhr may be the largest and most complete transition of peatland to intertidal, but it has rivals for quality, most notably at Kentra Bay to the north of Ardnamurchan. Kentra enjoys a highly

A peatberg stranded on Crinan's intertidal sands.

oceanic climate, mild and relentlessly wet. The bog formation here is technically categorised as eccentric, a predominantly Scandinavian type where the intricate pattern of pools appears to run contrary to the slope of the land. That gentle gradient can be followed down to the bay where the peat topples into the intertidal and the sandy shore.

Kentra supports one of the largest saltmarshes of the Highlands; the mouth of the bay is constricted by a rocky passage fortuitously aligned with Eigg and Rùm which moderate the reach of Atlantic breakers. Unusually for a Highland marsh there are muddy components amongst the sediments supporting pioneer glasswort communities as well as submerged banks of eelgrass. Otherwise the components are typical of the region, although the abundance of crofters' sheep suppresses more palatable species in favour of lawns of Thrift. That tight grazing of the sandier shores generates the bare ground and open swards required by the Seaside Centaury *Centaurium littorale*. This pink-flowered member of the gentian family grows here at the north-west limit of its British distribution, the most northerly populations being on the east coast on the shores of Loch Fleet.

'Neòinean Cladaich' Thrift at Kentra Bay.

Seaside Centaury.

Botanists are drawn to Kentra by the bog. The mildness of the climate supports species more commonly associated with the New Forest on the south coast of England; these include Brown Beak-sedge *Rhynchospora fusca*, Bog-sedge *Carex limosa* and the Oblong-leaved Sundew *Drosera intermedia*. The climate is wet but not quite wet enough to support a specialist of the adjacent Ardnamurchan peninsula. This is Pipewort *Eriocaulon aquaticum*, an emergent aquatic of peaty pools distinguished in flower by looking like a button-headed green straw. Pipewort is principally a North American species with European outposts on the wettest western peninsulas of the Highlands and Éire. The British populations are different at a genetic level from those of eastern North America, with only half the chromosomes. Such a difference suggests the plant is a long-standing component of our flora and not some recent colonist or introduction. The proximity of known populations to Kentra encourages further searching.

The advantage of attracting botanical expertise to a site is that groups of plants that are notoriously difficult to identify to species level will get studied; at Kentra these difficult plants include the eyebrights. Upper saltmarshes in the Highlands support some species of eyebrights that are known from nowhere else in the world. Relatively sheltered sandy saltmarshes subject to close grazing by sheep provide for one such species, the demure *Euphrasia heslop-harrisonii*. John Heslop Harrison, in whose honour the species was named, was Professor of Botany at Newcastle-upon-Tyne; he was a highly talented observer with much work of enduring importance. By all accounts Heslop Harrison was not an easy man to work with, being variously described as vigorous, resolute of opinion and dogmatic.

In 1934 Heslop Harrison embarked on a study of the flora of the Hebrides and formulated a theory that some islands had escaped the climatic extremes of the last glaciations. This theory was supported by a remarkable sequence of discoveries on Rùm of plants previously unknown to Britain. At that time Rùm was in private hands with restricted access, the privileged visitors including an annual party under the tutelage of Heslop Harrison.

In 1948 John Raven, a young Classics don from Cambridge, was given permission to visit the island, his academic background seemingly benign to Heslop Harrison's interests. Educated as a classicist, Raven was also a fine amateur naturalist, his appreciation of highland flora having been refined on his family's estate at Ardtornish. Raven found the rarities reported by Heslop Harrison but their habitats and plant associations were entirely wrong; the 'discoveries' were not scientifically credible. The matter was treated with courtesy and discretion; John Raven wrote a report for Trinity College which was lodged in their files until the death of the key parties. Raven concluded that Heslop Harrison had been deliberately indulging in the most culpable dishonesty. There was to be no public disgrace but the botanical records were quietly expunged. In a letter to the journal *Nature* in 1949 John Raven charitably suggested that Heslop Harrison's ardent and competitive personality may have laid him open to students' practical jokes.

Reassembling fragments

There is a risk in overemphasising the similarities between saltmarshes from distinctly different parts of the British Isles. Highland saltmarshes have their own character and suite of associates. The flora of the Highlands, well away from the reach of the tide, emphasise that the distribution of saltmarsh species can be a manifestation of complex environmental stresses and not just a simple association with saltwater.

The range of transitions from Highland saltmarshes to other habitats is far richer than can be dealt with in detail here. It is possible to have one's boots immersed at high tide whilst touching the Arctic-Alpine flora of Durness. Here Mountain Avens and Hair Sedge *Carex capillaris* can be found amongst strandline seaweeds. The boundaries between saltmarshes and sparsely vegetated beaches are diverse and at times ambiguous. Early Orache *Atriplex praecox* grows low in such intertidal zones but is not always grouped with saltmarsh species, unlike its close relatives the Spear-leaved Orache *Atriplex prostrata* and Long-stalked Orache *Atriplex longipes*. In their search to classify and categorise all living things ecologists are at risk of creating artificial boundaries in inherently complex ecosystems.

In most cases Highland saltmarshes are minority components of much larger landscapes. The extensive nature of land use in much of the Highlands has assisted the survival of transitions between habitats which survive, if at all, as fragments in the lowlands. Whilst it is unwise to exaggerate analogies between highland and lowland coasts, if you strip these landscapes down to their simplest components there are common features to be found.

There are many places around the coast of Britain where peatlands have formed in the headwaters of tidal rivers. There are extensive wetland systems of this nature in Broadland (Norfolk, Suffolk), the Humberhead Levels (Lincolnshire, Yorkshire) and the Somerset

Mountain Avens by the Kyle of Durness.

Levels. These English wetlands differ from Mòine Mòhr in that they have been extensively drained and their once-open landscapes have been broken up into fields. It does not take too great a leap of imagination to substitute one solid geology with another or to envisage alternative overlying patterns of fields and settlements. Standing on the bluff of Knapdale looking out over the Mòhr is reminiscent of the view over the Somerset Levels from the Polden Hills. The islands in the Mòhr at Dunadd and Barsloisnoch can be readily translated into Dundon or Glastonbury. The key elements are the same; they are all high ground in estuarine wetlands of saltmarshes and peat. What differs is the degree to which the land has been modified from its natural state. Mòine Mòhr gives insights into how other coastal wetlands have looked within recent ecological history; the Mòhr helps us interpret those relicts and inspires us to consider alternative futures for Britain's coastline.

Mòine Mòhr from the Knapdale hills.

Atlantic gateway

T he exposed extremities of Europe's Atlantic seaboard sustain one of the most distinctive landscapes of Britain, the machair. Within the machair is a complex of wetlands exposed to varying degrees of salt depending on the tide and season. The intertidal saltmarshes and brackish lochs of the Outer Hebrides are a small part of a much richer landscape reflecting the traditions of Gaelic culture. Extensive pastoral farming by the islands' crofters is integral to safeguarding the finest expressions of brackish aquatic habitats in Britain.

The Outer Hebrides form a 200km-long archipelago separated from the Scottish mainland by the deep waters of The Minch. Globally rare habitats derived from windblown shell-sands dominate the islands' machair landscapes. Machair is most fully developed in the central part of the island chain running from Eirisgeigh/Eriskay northwards to Beàrnaraigh/Berneray by way of Uibhist a Deas/South Uist, Beinn na Faoghla/Benbecula and Uibhist a Tuath/North Uist. Bilingual announcements on the ferries across The Minch herald what should be expected on arrival.

Saltmarshes are present on the islands wherever there is sufficient shelter to support vegetation in the intertidal. The principal components are similar to neighbouring mainland marshes apart from the abundance of the 'Luachair Bhailtigeach' Baltic Rush *Juncus balticus*. This is a species with a world distribution taking in the Atlantic coast of New England and Canada as well as the brackish marshes of the Baltic Sea. On the saltmarshes of the Hebridean Atlantic coast Baltic Rush contributes to the hues of the fine rush swards, its glaucous

OPPOSITE PAGE:
'Brisgean' Silverweed blooming in the machair.

'Luachair Bhailtigeach'
Baltic Rush.

stems complementing the duller green of the commoner Saltmarsh Rush *Juncus gerardii*.

Brackish conditions as reflected in the presence of Baltic Rush are a consistent feature of the islands' wetlands. Salinity is determined by the relative abundance of salt and water; on the islands those elements are in a constant state of flux. Extreme events, such as the storm of January 2005, can drive saltwater across the low-lying coastal plain. Sustained high wind speeds over much of the year provide for the distribution of fine salt particles. Winds over 10 knots, being stronger than a moderate breeze, occur on average every other day, with stormy months running half the year from October through to March. When gusts exceed 30 knots the normal salt-laden winds are supplemented with spume and spindrift. Rainfall counteracts the effects of salinity. In an average year Uibhist a Deas enjoys 202 wet days with an average annual precipitation at a little over 1.1m; this is not exceptionally high rainfall for the Hebrides but it is consistent in its application. All the habitats of the islands' lowlands are subjected to the competing forces of salt intrusion and freshwater dilution.

At Loch Phaibeil on Uibhist a Tuath intertidal marshes have formed in the comparative shelter of an opening in the dunes. Interspersed amongst the predictable cast of Highland saltmarsh species are numerous freshwater components including 'Lus na Peighinn' Marsh Pennywort *Hydrocotyle vulgaris*, 'Lus Riabhach' Marsh Lousewort *Pedicularis palustris* and 'Mogairlean' marsh orchids *Dactylorhiza* spp. The constancy of freshwater through rain and seepage is sufficient to maintain fen communities despite regular tidal inundation. One of the features of the upper marsh is a flood pasture characterised by 'Fioran' Creeping Bent *Agrostis stolonifera* and 'Brisgean' Silverweed *Potentilla anserina*. This is a widely distributed plant association found throughout Britain ranging from prestigious nature reserves to the muddy penumbra of kick-about goal mouths.

Greylag Geese *Anser anser* bring their young down to graze on the marshes at Phaibeil where their concentrated efforts create

RIGHT: Silverweed on shingle.

BELOW RIGHT: Greylag Geese produce a tightly grazed sward.

close-cropped swards of Silverweed; the association of goose and wildflower is related through their respective scientific names. The Latin *Anser anser* translates as 'Goose goose'. Greylag Geese are the wild creatures from which we, and presumably people of antiquity, bred the domesticated bird. *Potentilla anserina* in turn contains a specific reference pertaining to geese preceded by the genus *Potentilla* meaning a small powerful thing. *Potentilla anserina* earned its name for its robustness, a superficially slight plant capable of withstanding the onslaught of geese.

'Brisgean' translates into English as 'Brittle One' which at first seems contradictory to the Latin epithet. This brittleness is apparent should you try to dig the plant out. Brisgean grows from a mass of swollen roots that are frustratingly fragile. No matter how diligent the weeding there will always be fragments left to grow on. It is the quality of the roots that is recalled in the proverb

Brisgean beannaichte earraich
Seachdamh aran a' Ghàidheil

The blest silverweed of spring
A seventh part of the Gaels' bread.

Brisgean was blessed because in times of crisis its roots could be eaten and so help carry the people through to the next harvest. Margaret Bennett (1991) describes its starchy qualities being suited for grinding into meal for gruel or into flour for bread. Memory has it that Brisgean fed the islanders when potatoes were blighted and through dispossession at the time of the clearances. Some accounts suggest that Brisgean was cultivated before the adoption of potatoes and was grown to states of luxuriance far in excess of the downtrodden plants of the wild. Today Brisgean is a commonplace herb of the islands including in the fallows amongst the foundlings of potato beds.

The machair

As a preliminary to exploring the aquatic life of the machair lochs it is useful to understand their geography. At its simplest the machair is a landform generated by windblown shelly sand. On a bright summer's day its influence can be seen stretching out into the Atlantic, the startling white of the beaches contrasting with a shallow azure sea. Hebridean postcards promote the purity of such scenes; they are glorious but are far from being the only reality. Where favoured by wind and tide, the white sand strandlines accumulate deep beds of seaweed, great ribbons of kelp *Laminaria* spp. and assorted other fucoids broken off by winter storms. Rolled up with the decaying algae are the flotsam and jetsam of the ocean, broken fishing gear, the occasional dead seal and wrecks of eelgrass *Zostera* spp. beds. By such reeking tangles the abundant fertility of the sea is carried to the shore.

Machair beaches are aligned to rocky headlands and backed by high dunes. From afar the whiteness of their near-vertical faces appears

OPPOSITE PAGE TOP:
Machair shore.

OPPOSITE PAGE BOTTOM:
A machair loch.

Blackland farming.

architectural; south of Phaibeil the uninhabited Eilean Chirceboist rises from the intertidal like a citadel, its fortress arms embracing the marshes in its lee. Such freestanding islands are exceptions, for in most cases the dunes are fully connected to the mainland through plains of drifting sand. Where these plains meet the hills the calcareous sands and acid peats mix into biddable soils; this is where most people live, in amongst the fertile fields of the Blacklands.

Water permeates the convergence of rock and sand, forming a dendritic maze of sea lochs and bays, lochs, lochans and peat pits, all interconnected through elemental forces of tide, wind and rain. Environmental variations in the open waters of the islands defy ready summary; each differs from the other and those differences change over time. Counterintuitively it is possible for the same loch to have different salinity regimes at the same time. Studies of Loch Bi in Uibhist a Deas by Howson *et al.* (2014) have demonstrated such a range, with one part of the Loch being almost fresh whilst another is full-strength seawater.

In addition to salinity, the coastal wetlands are exposed to varying degrees of acidity. The bedrock of Lewisean Gneiss generates acidic conditions contrasting with the highly calcareous shell sand of the machair. Some lochs have formed over the sand; others are only exposed to drifting material, the volume determined by local topography. This interplay of salinity and acidity in turn affects fertility. Given all these interacting factors, it is unsurprising that the hundreds of waterbodies of the islands support an outstanding diversity of aquatic life.

Over the years ecologists have been hesitant to include aquatics in descriptions of saltmarshes. Such reticence would be unthinkable in freshwater ecosystems as an account of a fen or bog would be considered incomplete if it did not consider the pools and channels. Aquatic habitats can be insignificant in saltmarshes particularly where the transitions to other habitats are truncated. In coastal wetlands such as in the machair the separation of 'terrestrial' parts of saltmarshes from the 'aquatic' is indefensibly artificial; both are elements of the same continuum.

Recording the aquatic wildlife of saline wetlands is challenging as high levels of expertise are required to identify organisms to species level. The Gaelic for water-milfoil *Myriophyllum* spp. distils the issue; 'Snàthainn Bhàthaidh' translates as drowned threads, a description that can equally be applied to many aquatic species.

Drowned threads of
'Snàthainn Bhàthaidh'
Water-milfoil.

Interesting pondweeds

The importance of the Outer Hebrides for aquatic flowering plants emerged from the studies of Professor Heslop Harrison. In his handbook of British pondweeds Chris Preston (1991) recounts the tensions when George Taylor, of the British Museum, and his co-worker J E Dandy borrowed Heslop Harrison's specimens to assist their revision of the genus *Potamogeton*. Unfortunately Dandy and Taylor published a paper on Hebridean pondweeds that drew on a range of material which, with hindsight, may have been more fulsome in recognising sources. The reaction of Heslop Harrison to the perceived discourtesy was disproportionate to a degree that even his sympathisers felt it extreme. There ensued a vitriolic public dispute in which the professor unwisely challenged his rivals on technical matters of nomenclature. Whilst the world was at war, academics disputed pondweeds. The *coup de grâce* fell in 1944 with Dandy and Taylor's riposte in the *Journal of Botany*, an indisputable scientific case presented with withering sarcasm. Heslop Harrison's revenge was refusal to permit any of his associates to pass further material to Dandy or Taylor; the matter ended in bitter recriminations with important records unvalidated and the scientific record incomplete. More recently, less volatile botanists re-recorded some of Heslop Harrison's unconfirmed species, so vindicating the contribution he and his team made to understanding the flora of the Outer Hebrides.

There are currently 18 species of *Potamogeton* pondweeds and their hybrids known from the Outer Hebrides with a good range present within, and demonstrably tolerant of, brackish lochs. Unsurprisingly, there are no local names for any of the pondweeds of the coastal margin. The English names are mostly concoctions reflecting the fashion that scientific terms are too exclusive. Anyone developing an interest in the botany of *Potamogetons* will be long past worrying about such matters.

The pondweed that grows in the strongest saline solutions is the Fennel Pondweed *Potamogeton pectinatus*. As its name suggests, it looks like fennel insomuch as its leaves are little more than flattened hairs. Fennel Pondweed and the even more filigree Lesser Pondweed *Potamogeton pusillus* can be abundant throughout the machair lochs yet are a great rarity across the remainder of northern Scotland.

The Fennel and Lesser Pondweeds appear positively bulky compared to the Shetland Pondweed *Potamogeton rutilus*. In this species few individual leaves achieve a millimetre in girth, many much less in stature. This slender greenery can extend up to 80 times as long as it is wide. I once had a brief opportunity to admire plants growing near the RSPB's visitor centre at Balranald. Between the access road and the shore Shetland Pondweed is found in the interconnected lochs of Grogary and Croghearraidh. The site lies at sea level with a small stream to the ocean and so is exposed to moderate amounts of salt. My misfortune was that a Short-eared Owl *Asio flammeus* was working the valley, disturbing the waders to the delight of gentlemen with 'scopes. My life would not have been worth living if I had put up their prey.

I had better fortune walking into Loch Bhornais. A firm bed of white sand was coated in young stoneworts out of which rose the Slender-leaved Pondweed *Potamogeton filiformis*. I have yet to see an illustration which does service to its charm. Rather than a tuft of clingy foliage there was elegance in its stems, every leaf poised in a buoyant open gesture square to the light. The clarity of the water invited a sip; there was just a little salt, not at all briny, a pinch slight enough to cook spinach.

The aquatic flora of Loch Bhornais.

Rarest of all the Hebridean aquatics is *Potamogeton epihydrus*, the American Pondweed. My delight in first finding this plant was disproportionate to the effort required to see it. Once on Uibhist a Deas it was a simple matter of following the published accounts to Loch Ceann a' Bhaigh at the tidal headwaters of Loch Aineort. An amble across the bog took me to a shallow bay where the obscure object of my desire was rooted in a suspension of loose peat.

It being early summer the pondweed was still fully aquatic; its elliptical floating leaves had yet to develop. The submerged foliage comprised short ribbons of the finest translucent fettucine with crisp undulating margins. Each leaf appeared as if infused in fruit tea; there were delicate tannins with a hint of hibiscus. From the tip of each ribbon a central vein ran towards the stem gaining parallel companions until multiple pale bands formed a flattened fascine; it was the most beautiful thing.

The natural distribution of American Pondweed in Europe is confined to a few lochs and peaty pools between Loch Cean a' Bhaigh and the interlinked Lochs of Olaidh. These lochs lie just above the normal tidal limit; Loch Cean a' Bhaigh is connected to Loch Aineort by way of a channel with a rock sill and so is open to tidal surges from the east and sea spray from the west. The flowering plants growing with the pondweed were consistent with freshwater but the accompanying Rough Stonewort *Chara aspera* is tolerant of the weakly brackish conditions. Having walked the site I could not discount salinity from affecting the ecosystem, but there is nothing to assert that such influence was anything but transitory.

On the same day as visiting Loch Cean a' Bhaigh I walked the machair of Bornais township to the headland of Rubha Àird a' Mhuile, the most westerly point on the island. Perched just above the tideline is another brackish loch from which emerges the ruined broch of Dùn Vùlan. Brochs are Iron Age multistorey round houses typically constructed on small islands; they are high-status buildings in prestigious and defendable locations.

In the early 1990s archaeologists from the University of Sheffield excavated Dùn Vùlan, the site being at risk from coastal erosion. Amongst their many finds they recovered parts of the original roof. At the time of construction the builders of the broch would have had difficulty accessing timber; two thousand years ago there were no natural forests on the islands and certainly no plantations. It was possible to import material, at great expense, but there were other solutions. Forensic examination of the timber fragments at Dùn Vùlan determined that the wood was larch *Larix* sp., an interesting occurrence given the nearest stands at that time were either *Larix decidua* in the Alps or *Larix sibirica* in what is now Russia. Investigations to a cellular level concluded the most likely candidate was *Larix laricina*, the Tamarack of the north-eastern seaboard of America.

Fortunately the wetlands surrounding the broch not only preserved fragments of roof but also the debris from where the carpenters prepared the beams. The discarded chips were riddled with *Teredo navalis*, Shipworm, a burrowing saltwater clam notorious for its ability to weaken timber. The shipworm indicated the timbers had been a long time at sea from which Taylor (1999) deduced the broch had been roofed with driftwood whose origins were a continent away.

The carpenters working on Dùn Vùlan were not alone in using flotsam. A collation of records by Dickson (1992) of similar timbers

recovered from around the Hebrides and Northern Isles indicates that spruce trees *Picea* spp. were more commonly used; spruces grow both in Scandinavia and North America. The archaeological record is evidence that ocean currents are capable of bringing the driftwood of American coastal forests across the Atlantic. As timbers make that passage it is probable that smaller fragments of vegetation have done likewise. Where seeds and rhizomes are tolerant of saltwater they may make landfall alive and by chance find somewhere to grow. The archaeological record offers a possible, even probable, mechanism for species such as American Pondweed to find a footfall on the shores of Europe.

Driftwood on a forested Newfoundland shore.

Lagoon denizens

To the casual eye stoneworts look like aquatic plants. For generations botanists have appreciated their contribution to wetlands of exceptional quality and have been comfortable recording them as if they were flowering plants. However, stoneworts are not plants, they are algae, which are more closely related to the tangles of wrack on the shore than to any pondweed. Compared to the simplicity of kelp fronds, the structure of stoneworts is highly developed, their stems and leaf-like whorls reminiscent of the drowned threads of water-milfoils. Early botanists did not fully understand their otherness, with John

Ray (1690) regarding them as horsetails *Equisetum* spp. and Thomas Johnson in his revision of *Gerard's Herbal* (1633) classifying them amongst mare's-tails *Hippuris* spp.

Stoneworts or, to the technically minded, charophytes get their name from encrustations of calcium which gives them a gritty texture together with a degree of rigidity when taken from the water. Material in the hand allows the appreciation of their unwholesome scent which has a quality akin to chlorinated garlic.

From an ecological perspective stoneworts are of interest for what they tell you as much as for what they are. Each species has its own range of environmental sensitivities and so their presence gives insights into water chemistry; Nick Stewart (1996) has called them the connoisseurs of clean water. That clean water may be salty; indeed there are indications that the evolutionary history of stoneworts took them from being marine algae into brackish and subsequently freshwater conditions from where they contributed to the evolution of life on land. Compared with the usual soft tissues of algae, the reproductive structures of stoneworts are robust and so are readily fossilised. Charophytes were definitely living at the time of the Jurassic dinosaurs, with other fossils suggesting far greater antiquity into the Silurian era, 400 million years ago.

Of the saltwater stoneworts it is the Foxtail Stonewort *Lamprothamnium papulosum* that is described by Jenny Moore (1986) as growing in the widest range of saline conditions from almost full-strength through to quarter-strength seawater. *Lamprothamnium* was first discovered on the Outer Hebrides in the 1980s and is currently known from over a dozen lochs. There is a cluster of sites on Uibhist a Tuath by Clachan an Luib where a short stretch of bog forms the watershed between The Minch and the Atlantic. Here the basin of Oban na Curra is connected at high water to the tidal Loch Euphort. Foxtail Stonewort is the only charophyte growing in this bay where freshwater inputs are limited to rainfall. A few hundred metres to the east is Oban nam Fiadh, another tidal reach of Loch Euphort, which grades upstream into the multiple freshwater arms of Loch Carabhat. The broader range of salinities found here is suitable for both the Foxtail Stonewort and the Rough Stonewort *Chara aspera*. Rough Stonewort is catholic in its tolerances, being capable of growing in brackish water particularly if shell sand reduces the acidity. It was Rough Stonewort that accompanied my appreciation of the American and Slender-leaved Pondweeds.

Immediately to the west of Oban na Curra across the watershed is Oban a' Chlachain which enters the Atlantic in the lee of Baile Sear. Baile Sear is an island served by a causeway where a series of eight small lochs is set within machair sands and Blacklands. The lochs are connected to the sea by way of ditches and brackish marshes and so are open to occasional storm surges together with the constant effects of salt spray.

The assemblage of stoneworts on Baile Sear is unrivalled. Rough Stonewort is in six of the lochs along with eight other species including Britain's most secure population of Bearded Stonewort *Chara canescens*. One of the difficulties in understanding extreme rarities like Bearded Stonewort is that it may not grow in enough sites to exhibit its full range of environmental tolerances. A solitary population may be in an optimum habitat or could be on the verge of dying out; without comparators it is difficult to know. In north-western Europe, Bearded Stonewort is found in a range of circumstances from the mildly brackish Baltic to inland salt springs. In Britain it has been known from numerous sites, mostly brackish, but also from freshwater ponds in Suffolk. Apart from Baile Sear its other substantial British population is found in a series of worked-out clay pits at Orton near Peterborough in the English Midlands. The clay extracted for brick making is of marine origin, with minerals weathered from quarried exposures contributing to dilute brackish conditions. The water chemistry is reflected in the vegetation, with both the Fennel and Lesser Pondweeds growing amongst Bearded Stonewort. Maintaining the excellence of water quality required by Stoneworts is a more likely prospect at Baile Sear than in an old brickworks which is steadily becoming incorporated into one of Britain's fastest-growing cities.

Bearded Stonewort.

When visiting any site it's worth taking time to consider what you have not seen. Foxtail Stonewort has not been recorded at Baile Sear, but then the salt concentrations are less than in neighbouring sites. There is however one species that is unexpectedly absent, the Baltic Stonewort *Chara baltica*. Self-evidently the Baltic Stonewort is associated with the Baltic Sea as well as being widely distributed around the coast of north-west Europe. Local to Baile Sear, this stonewort is found in numerous inlets around the headwaters of the nearby Loch nam Madadh, in some of which it is accompanied by Rough Stonewort and Foxtail Stonewort. In Loch Bi Foxtail Stonewort grows in the more saline Atlantic reaches and is quite separate from the Baltic and Rough Stonewort beds of mildly brackish waters. The absence of Baltic Stonewort records from Baile Sear could possibly be due to chance; equally it may be present but has yet to be identified. Alternatively our knowledge of the tolerances of stoneworts is imperfect and there could be some unknown factor affecting distribution that has yet to be appreciated.

Pondweeds and stoneworts are useful indicators of highly specialised habitats that support species requiring even greater expertise to identify. Mud-snails *Hydrobia* spp. are commonly found in the mudflats and creeks of saltmarshes. The snails that are most abundant in the brackish water of Loch Bi are *Hydrobia acuta neglecta*; their shells range in size from 1–4mm and are found in abundances of up to 300,000 animals to a square metre. Even microscopic examination of their shells is not always definitive in identifying individuals to species level. In recent years specimens have been determined by Chevalier *et al.* (2014) using DNA analysis, a far cry from the traditional skills of the field naturalist.

Not all lagoon specialists are as cryptic as mud-snails; the Lagoon Cockle *Cerastoderma glaucum* grows to 5cm and is identifiable, with practice, from its shells. Lagoon Cockles are known from around the British coast in lagoons as well as in estuaries with a strong input of freshwater. The adult cockles are relatively robust, unlike the younger life-stages of planktonic larvae and freshly settled spat. Their vulnerability to wave action and desiccation over the summer breeding season restricts the cockles to relatively sheltered conditions as are found in Loch Bi and in the headwaters of Loch nam Madadh.

Reduced salinity in a machair loch does not necessarily exclude what are generally regarded as marine species. Not all seaweeds require full-strength seawater; for example, the Horned Wrack *Fucus ceranoides* thrives in brackish lagoons and on rocky foreshores flushed with freshwater. Eelgrass beds *Zostera* spp. grow with tasselweeds *Ruppia* spp. in distinctly brackish conditions of half-strength seawater. Tasselweeds are closely related to *Potamogeton* pondweeds but are associated with much higher concentrations of salt than the most tolerant Fennel Pondweed.

The literature and vocabulary of brackish water habitats tend to distinguish lagoons from brackish habitats with open connections to the sea. Lagoons are generally regarded to be saline waterbodies detached from the intertidal, a convenient but artificial division. It has been estimated that there are about 3000ha of saline lagoons in Britain. Scotland supports twice as much of this habitat as the other two countries combined. The lochs of the Outer Hebrides represent a significant proportion of the nations' resource, Loch Bi alone accounting for over 800ha.

Brackish lochs and other coastal wetlands in the Outer Hebrides are not ecologically immaculate; many have been modified to some degree but none to the point of destruction. The scale and diversity of the wetlands demonstrate the subtle expressions of enhanced salinity in aquatic habitats. The exemplary quality of these wetlands is maintained through the modern manifestations of the human traditions of the machair.

Farming the machair

Machair is a geographical feature that is synonymous with the communal farming practices of the islands' townships. A township is a community of crofters, farming families whose tenancies are secured to guard against repetition of the historical injustices of the Highland clearances. A croft usually comprises a family home and farm buildings together with permanently enclosed fields under the sole management of that crofter. It is useful to have such land to make hay, to raise livestock close to home or to grow crops for the household.

As well as their fields, crofters have allocations of rights to use the machair which contributes both grazing land and winter feed to the local livestock economy. The combined herds of each township are free to graze across the machair from late summer through to spring, the season being determined by weather and the degree of winter flooding. There are practical advantages in the plains being saturated in winter; water helps protect the sand from the scouring effects of the gales.

In spring, on a day appointed by the township, the cattle are driven off and crofters with the appropriate rights are then free to plough their allocation and establish spring crops. These are usually cereals, not for human consumption but to supplement the diet of the cattle through the winter. The scale of these arable fields is substantial; it is ploughland of an order parallel to anywhere else in Britain, but on this point the similarities with high-input modern farming end.

The soils of the machair are hungry, the crops are thin and there is an abundance of bare ground and 'weeds'. The growing season may

Ploughland by a brackish loch.

be short but at this latitude day length is long so the usual crops are black oats and rye. There is still some cultivation of an ancient form of barley with six rows of florets rather than the conventional two. This Bere barley responds well to local conditions so it is often one of the first crops to be gathered in.

In every year there are fallows amongst the ploughland, allocations left to rest for a season or two to rebuild natural fertility. In some years substantial proportions of a machair may be fallowed which, with judicious fencing, provide summer pastures for the livestock. Traditionally, plots were mulched with tangles of seaweed harvested from the beach, a practice still apparent in potato patches amongst the dunes. The arable season ends in late summer when, on another appointed day, the cattle are let back onto the machair – a powerful incentive to ensure a timely harvest.

Saltmarshes and lochs embedded in the machair are inadvertent beneficiaries of this farming regime. The marshes remain grazed, the fine turf and nuanced transitions into brackish and fresh thereby secured against domination by coarser species. This grazing extends to the sandy bottoms and margins of the lochs where a little disturbance by trampling feet creates opportunities for adventitious aquatics.

Water quality in the lochs is safeguarded by the extensive nature of the husbandry. Natural fertility is scarce and artificial fertilisers are expensive to import so surpluses, which could drain off into the lochs, are not applied. Similarly, herbicides and pesticides are conspicuous by their absence. There is sufficient land to cultivate to secure enough feed for the cattle without having to intensify.

The farming practices of the machair contain elements that are familiar to any student of pre-enclosure Britain. The worked landscape contains portions that are farmed both individually and collectively. The emphasis is to secure the family and their livestock through the lean months; arable cropping is part of this subsistence strategy rather than farming for a surplus. The success of the township depends on communal enterprise between individuals. Such historical elements don't detract from the practical modernity of the landscape; the machair is a place of new tractors, barbed wire and silage wrap. The superlative wildlife of the islands arises from modern people finding contemporary ways of farming that respect the traditions of a rich cultural heritage.

South Uist, a lived-in landscape.

The merse of the Solway Firth

The Solway is a large and predominantly rural complex of estuaries containing the most extensive and interconnected saltmarshes in Scotland. A journey along the face of a marsh will take you through eroding cliffs and onto silt banks freshly colonised by saltmarsh grasses. Much of the Solway, along both nations' shores, is unmodified through engineering, making it one of the least altered large estuaries in Britain. The Solway is exceptional for the extensive areas of upper marsh where saline influences grade subtly into freshwater and terrestrial habitats.

Scotland and England are separated by the Solway Firth, an estuary that defines the political and linguistic boundary between nations. Along this frontier are monuments to warring factions and outposts of empire. It was on Burgh Marsh, amongst his army encamped on the English shore, that King Edward died in 1307, the Hammer of the Scots succumbing to the indignities of dysentery. The linguistic divide is still manifest in the naming of saltmarshes, 'merse' to the north, 'marsh' to the south, both terms originating from the same Old English *mersc*. In contrast it is the Old Norse *fjorthr* from which the estuary takes its title Firth. A *fjorthr* is a narrowing of the sea and is widely used around Scotland to name both estuaries and inshore sea lanes.

Agriculture has long been the principal industry of the coastal fringe on both sides of the Solway. The livestock sector is prominent and most of the saltmarshes are, to varying degrees, grazed by cattle and sheep. In common with many marshes formed over sandy substrates, the merse of the Solway are characterised by extensive grassy flats crossed by incised freshwater streams. Where regularly grazed, the

Memorial to Edward at Burgh.

OPPOSITE PAGE:
The merse beyond Caerlaverock Castle.

merse are dominated by fine grasses. Herbaceous flowering plants are concentrated around freshwater flushes and in freely draining areas along creek edges and on higher terraces. This character is inherent in the soils and natural drainage of the merse. The removal of grazing does not result in more flowery turf but in a rank mat of coarse grasses.

There is a remarkable completeness and continuity to the saltmarsh habitats along the Solway. Landward limits of the merse are clear where the intertidal abuts the steeply rising ground of dunes, cliffs and hills. More often the merse grades through less frequently flooded land which may be conventionally farmed and divided into fields. Despite the orderly presentation of banks and boundaries, in most years the highest tides drive saltwater far inland.

Barnacle Geese

Between Southwick Burn and the mouth of the Annan the north shore of the Firth is managed as a contiguous series of nature reserves by a partnership of conservation bodies and the local landed estates. Between them they have refined the grazing of the merse so that when the cattle move to their winter quarters the sward is in an ideal

OPPOSITE PAGE TOP:
Burgh Marsh, Cumbria.

OPPOSITE PAGE BOTTOM:
Mersehead, Dumfries.

Solway dawn, December 2015.

Barnacle Geese at
Caerlaverock.

condition for grazing waterfowl. Wintering flocks find this coastline
particularly attractive where undisturbed roosts on estuarine sandbars
are found alongside the nutritious fine grasses and clovers of the merse.

My first encounter with the Solway's Barnacle Geese *Branta
leucopsis* was in late December 2015, hunkered beneath a thin hedge
at Caerlaverock, waiting for dawn. If there was a horizon that day it
was lost in low cloud and driving rain. I had taken the field path to
the merse in the hope of witnessing the passage of geese from roost
to feeding ground. Every surface, in every direction, was awash and
running with mud. Dawn was undetectable beyond a hint of contrast
on the sodden fields. With that relief the geese arrived, conversationally
calling to each other through the murk. The flocks dropped gracefully
out of the gloom onto the flooded furrows.

By daylight I'd seen more birds from my vantage point than had
been their entire population in my youth. In the late 1940s the
population of Barnacle Geese on the Solway stood at a few hundred
birds. Their decline from 10,000 geese in the late 19th century has
been attributed to hunting and disturbance. It is self-evident that
hunting is fatal to the few birds that are shot; more importantly, the
associated disturbance expends energy in otherwise unnecessary flight
as well as restricting the time when the geese can feed. The net results
are geese less fit to migrate and, on arrival, less successful in breeding.

A well-fed bird arrives earlier than the majority of the flock, occupies optimal breeding sites, secures a stronger mate, fledges larger clutches and has time to feed up ready for the long haul south.

From the 1950s the Duke of Norfolk's estate at Caerlaverock has been the focus of a visionary approach to conservation that promotes the co-existence of geese with wildfowling, farming and fishing. That vision drove forward one of the great conservation success stories.

By the 1960s the combination of regulated wildfowling and the provision of refuges enabled the population of Barnacle Geese on the Solway to rise to around 5,000. Through the capture and ringing of birds in Scotland and Norway it was established that the Solway was the wintering ground of virtually the entire population of Barnacle Geese nesting on the Svalbard archipelago in the Norwegian Arctic. Depending on the choice of route, the distance between the wintering and breeding grounds is between 2,500 and 3,000km. This twice-yearly migration is undertaken in a series of flights with traditional staging posts along the Norwegian coast and on Bear Island, south of Svalbard. In favourable conditions the migration can be made in just five days but foul weather can extend the journey to over a month.

Breeding grounds for Barnacle Geese on Svalbard.

The combined flocks of Barnacle Geese wintering on the Solway have risen to well over 30,000 birds and that growth currently shows no sign of levelling off. There are however natural limits to any population. In recent years Svalbard's Polar Bears *Ursus maritimus* have taken to predating nest sites. Should Barnacle Geese become increasingly favoured as prey this could restrict further growth in the population. There are also risks with the wintering quarters being limited to the Solway. A local catastrophe, such as the outbreak of disease or a build-up of parasites, could curtail the value of their wintering grounds.

Over time the breeding and wintering grounds of waterfowl change to reflect shifts in the climate and other environmental conditions. Before flying north some of the geese gather on the marshes on the English side of the inner Solway at Rockcliffe. Since the 1990s a proportion of these flocks has delayed leaving the Solway and has reduced the time they spend at staging posts on the Norwegian coast. The Solway/Svalbard birds are one of three world populations of Barnacle Geese, the other two breeding in eastern Greenland and the Siberian archipelago of Novaya Zemlya. In recent decades both of these populations have started to diversify their breeding grounds and by doing so have reduced the distances they need to migrate. It is possible that, in the future, alternative wintering grounds may become adopted by the flocks. In the meantime the arrival of Barnacle Geese on the Solway remains one of the most cheering wildlife spectacles of Britain.

Ultimate survivors

The landward boundaries of the merse are exposed to a particular suite of environmental stresses. To live in this zone an organism needs to be able to cope with combinations of drought and inundation, of great variations in salinity together with the constant depredations of livestock and geese. It is one thing to draw on the seasonal abundance of the marsh and another to seek to complete entire life cycles under such pressures. The combined stresses of the upper merse determine the range of species present and demand specialisation to survive. It is on the cusp of drowning and drought where some of the Solway's exceptional wildlife can be found.

In amongst the seasonally wet pools of the upper merse live Tadpole Shrimps, creatures of temporary waters with a life cycle attuned to

A fully grown
Tadpole Shrimp.

sporadic phases of wetting and drying. When their pool is dry their eggs lie dormant in the dust, a dormancy that can last for decades. On wetting, some eggs, but not all, hatch out and grow to adulthood in less than a month. If the pool dries before the hatchlings can lay eggs of their own then the residual stock remains, awaiting the next opportunity.

Tadpole Shrimps and their ancestors can be traced through the deep time of earth's geological history. The Tadpole Shrimp known from the Solway is *Triops cancriformis*. This species is preserved in the fossil record in exquisite detail; fossils of *Triops* are indistinguishable from the creatures that live in the merse today. The earliest rocks containing fossil *Triops* date from shortly after the world's most comprehensive mass extinction at the end of the Permian period. By the time the dinosaurs of popular imagination walked the earth *Triops cancriformis* had already been in occupation for over 130 million years. The shrimp went on to outlive those dinosaurs through the mass extinctions of the Cretaceous, the subsequent redistribution of the world's continents, the building of mountain ranges and the climatic extremes of the ice ages. *Triops* has remained unchanged as a species for at least 220 million years. Some academics have had the audacity to present its success as being evolutionarily stagnant. There are few pressures to evolve further on a creature that already has all the adaptations it needs to survive the upheavals of geological time.

The tolerance of *Triops* to extreme environments has resulted in it being the dubious beneficiary of much experimentation. One wonders what respect is reflected in the scientific paper by Carlisle

(1968) '*Triops* (Entomostraca) eggs killed only by boiling'. There are many other studies by Hempel-Zawitkowska (1968, 1969, 1970, and Klekowski & Hempel-Zawitkowska 1968) where the adults and eggs have been exposed to extremes of light, dark, heating (both boiling and desiccation), freezing, oxygenation, stagnation, ultraviolet radiation, and opportunities to hatch in diluted seawater and unnatural solutions of a battery of salts KCl, $CaCl_2$, $MgCl_2$, K_2SO_4 and $MgSO_4$.

The *Triops* populations of the Caerlaverock merse are unquestionably the largest and most numerous in Britain. There is a scattering of historical records from across lowland England, mostly attributable to the peripheral habitats of heathland commons. These landscapes tend to have been lost to enclosure and conventional farming. Even where the physical extents of heathlands have survived, most are no longer grazed and their seasonal pools have been subsumed beneath rank grass and woodland. The exception is the New Forest, where a population still thrives.

I first saw *Triops cancriformis* in the New Forest in the mid-1980s. By chance the nearest pub to my lodgings was opposite the pond where the creature had been known since the 1930s. Through the generosity of Mr R E Hall, whose mantelpiece bore a portrait of *Triops*, I was initiated into the science of ephemeral waters and encouraged to undertake research of my own. My studies were confined to simple observations of the pond and its wildlife. There is something addictive about working with wild populations of *Triops*; they are fickle skulking creatures. You can revisit the same site dozens of times and never find them. Then, a snatched visit peering through February ice, or a glance into a foetid droughted stew, and there she is. It is always a she in Britain; in the northern latitudes all *Triops* are self-fertile females. I soon learned there are only two records you can make when surveying for *Triops*, namely: (a) *Triops* were present; and (b) I did not find *Triops* on this occasion.

Frank Balfour-Browne started out in life in a respectable profession. He qualified as a lawyer and was admitted to the bar in 1898. Subsequently, Balfour-Browne set aside legal practice to study zoology and embarked on a career of teaching and research culminating in his appointment to the chair of entomology at Imperial College, London. Since his schooldays he was fascinated with water beetles and spent a lifetime studying them. It was on such a search that in September 1907 he discovered *Triops* in two shallow grass pools on the Preston sea merse, near Southwick on the Solway's shore. In one of the pools *Triops* was so abundant that his net became half-full on a single sweep. Having unsuccessfully searched the surrounding area for more specimens, he returned to the original site a few days later to find that the local gulls had also discovered the mass of food with the edges of the pool covered by the inedible carapaces. He returned to the merse many times over the next 40 years but did not find any more specimens. Persistence has its rewards and in June 1948 at the age of 74 he revisited the mouth of the Southwick Burn. It was his son, who accompanied him on that visit, who was the first to re-find *Triops* and then together they explored a population extending over several pools. Balfour-Browne reported their discovery in a letter to the journal *Nature*, concluding his account with the question as to how many other pools around the Solway may support *Triops* undetected by naturalists.

The intertidal habitats of the Solway are highly dynamic. The merse is in a constant process of accretion and erosion; such changes are integral to the life of a saltmarsh. At Southwick the merse known to Balfour-Browne is now a part of the foreshore.

Balfour-Browne posed the question as to whether additional populations of *Triops* remained undetected, and after 50 years his observation proved sound. In 2004 *Triops cancriformis* was discovered some 19km to the east of Southwick in the upper reaches of the merse at Caerlaverock.

The discovery at Caerlaverock was serendipitous. Larry Griffin of the Wildfowl and Wetland Trust was surveying pools on the nature reserve looking for the tadpoles of late hatching Natterjack Toads *Epidalea calamita*. The merse pool in which he found *Triops* was unremarkable. It was not much more than a puddle, a trampled hollow where cattle rub themselves against a disused concrete fencepost.

Since 2004 considerable effort has been made to understand better the status of *Triops* on the Solway. As well as netting ponds for adults it is possible to hatch *Triops* in a bucket from mud samples gathered from across the merse. Many additional specimens have emerged from surveys ranging across the Caerlaverock nature reserves managed by both the Wildfowl and Wetland Trust and Scottish Natural Heritage. The shrimp is present in at least a dozen or so pools in fewer than a handful of sites. In wet seasons the standing waters of the merse flow into one another and challenge the notion of 'sites' necessary for biological recording. Some populations are refreshingly prosaic as all the shrimp needs is a temporary body of water; a vehicle rut or a poached gateway will do.

Triops is the ultimate survivor. They have survived across geological time because they can tolerate ordinarily intolerable levels of stress; to survive they need the continuity of stress.

Natterjacks

On first impression Natterjack Toads are paradoxical in their choice of habitats. In Britain they are species of lowland heaths, sand dunes, an English Lakeland fell and, on the Solway, the merse. The common factor is that all these landscapes are open with ephemeral waterbodies unshaded by trees. The vegetation is typically sparse with plentiful loose ground or debris into which the toads can take shelter. Natterjack Toads are species of early-stage succession vegetation communities; adults will forage under tall vegetation but must return to open ground to breed. On the merse these habitats are provided by the dynamism of the shoreline together with the effects of grazing animals.

Along the Solway shore, populations of Natterjacks can be found in a range of habitats, in pools in conventionally farmed fields and from there seaward into the merse. Many of their breeding ponds are distributed across the upper saltmarsh and are distinctly brackish. Throughout Europe there are populations of Natterjacks with varying degrees of tolerance to saltwater. As salinity rises over about 5ppt (parts per thousand) the development of tadpoles is increasingly adversely affected, with immersion in full-strength seawater proving fatal. The combined populations of the Solway have been estimated as accounting for as much as 20 per cent of the British population. This proportion changes over time as Natterjack populations go through periods of boom and bust reflecting the inherent ecological instability of their breeding grounds. The longevity of the adults, the abundance of spawn and the long breeding season are all adaptations that help maintain the continuity of populations through leaner years.

As Britain warmed after the last ice age its landscape became progressively amenable to plants and animals adapted to temperate climates. Natterjacks tolerate severe winters to a sufficient degree to have colonised the foothills of the Alps and into eastern Europe. They do however need warm summers to breed; consequently during the height of the ice age some 20,000 years ago they were confined to refugia in the Iberian peninsula. Genetic studies by Rowe *et al.* (2006) have described how, as the climate changed in their favour,

populations of Natterjack spread northwards colonising Europe from Iberia to the Baltic, with at least two populations spreading into what was to become Britain. The toads that colonised eastern England are genetically different from those that colonised the west, that western population reaching the edge of the European continent on Éire's Kerry coast. The sequence of colonisation is clear but the timetable is less so; by their very presence we know the toads succeeded in crossing into England before the opening of the North Sea and were one of the few herptiles to arrive in the island of Ireland.

The dispersal of Natterjacks across Europe and into the British Isles was accomplished in a few thousand years. In terms of simple logistics the toads are sufficiently mobile to crawl that distance in the time available. Over the course of a year an adult Natterjack will move hundreds of metres, and exceptionally over two kilometres, from its natal pool. We know from the behaviour of Natterjacks today that they are not loyal to where they were hatched and will take advantage of newly formed pools. For the colonisation of sites across Europe to succeed there will have been a geographical and temporal continuity of Natterjack habitats across the continent for thousands of years.

Male Natterjack calling for a mate.

Aurochs and Elk

It is not known precisely when, or by which routes, Natterjacks arrived in the Solway. On arrival they will have encountered a coastline already occupied by creatures that were either more mobile, having made a similar journey in a shorter time, or that were long established due to their tolerance of a broader range of environmental conditions. Between ten and twelve thousand years ago Britain's native fauna included large herbivores including Elk *Alces alces* and Aurochs *Bos primigenius*. It is probable that Natterjacks were amongst the species moving north and west as summer temperatures warmed and the climate stabilised. All of the larger mammals present in Britain at that time are known to exploit saltmarsh habitats elsewhere in Eurasia where their natural ranges extend to the coast. The early-stage succession landscapes required by Natterjacks would, as today, have been part of the succession of habitats of the merse. Similarly, as today, grazing animals would have played a role in maintaining those habitats throughout the merse and into its landward margins.

Derek Yalden's *History of British Mammals* (1999) summarises accounts of the remains of Elk and Aurochs from archaeological sites around the Solway. Elk remains, recovered from the River Cree, have been dated by radiocarbon assay from the Bronze Age, some 3,925 years ago. The Cree record suggests Elk survived on the Solway for longer than anywhere else in Britain. The remains of the Solway's Aurochs have not been dated with similar precision. The context of the local finds is Middle Stone Age but there are abundant records of Aurochs throughout Britain up until the Bronze Age.

Though long extinct in Britain, Elk are still widely distributed through the northern hemisphere and their ecology is well studied. The now-extinct Aurochs are the native cattle of Europe and for nearly ten thousand years after the last glaciations were the largest grazing animal in our landscape. Between them Elk and Aurochs contributed to the moulding of prehistoric Britain.

To what degree Elk modified the vegetation of the Solway coast for species such as Natterjack is open to speculation. For most of the year Elk are solitary creatures of gladed forests that are attracted to open country where there is aquatic vegetation and salt. The remains of Elk recovered from the Solway are from coastal wetlands and so it is probable they exploited, and by doing so helped maintain, the open habitats of the coast.

Elk are aquatic foragers.

Although the species is now extinct, it is possible to make informed judgements on how Aurochs fitted into the landscape and prehistoric ecosystem of the Solway. We know much about Aurochs, not least as the species survived into the modern era, the last cow dying in the Polish hunting grounds of Jaktorowska in 1627. Aurochs were herd animals; for most of the year cows and their young grazed independently from bachelor herds and solitary bulls. The diet of Aurochs was mostly grass and herbage, their preferred habitats being open grasslands, particularly amongst marshes and swamps. There is some evidence of browsing, particularly during lean seasons, when the herds would have taken advantage of low-growing shrubs and trees. Aurochs were significantly larger than their domesticated brethren; however, in their feeding and behaviour they differed little from modern free-ranging cattle.

Throughout the period following the latest ice age people have lived amongst wild animals and have modified their populations and behaviour. In the recent historical past human impacts have reduced the diversity of our native fauna; Elk and Aurochs were early extinctions associated with the adoption of settled agriculture from the New Stone Age onwards. Domesticated cattle appear in the archaeological record from some 5,500 years ago, 2,000 years before the extinction of Aurochs in Britain. Up until then it appears Aurochs were using coastal wetlands and shared that landscape with their domesticated relatives.

For millennia Britain's open habitats have developed in the presence of large grazing animals. There is continuity of grazing by large herbivores from the prehistoric herds to the domesticated animals of today. Shadows of prehistory may be seen where wild deer still come to the merse to graze together with the seasonal congregations of wildfowl. Natterjacks and other creatures of open habitats require such continuity to survive.

Through the Needle's Eye

In 1999 the distribution of Natterjack Toads on the Solway was supplemented through the introduction of a population into the RSPB's nature reserve at Mersehead. The toads have settled in to their new home and in spring a lusty chorus of amorous males serenades the shore. Within earshot of their calls the merse crosses the outfall of the Southwick Burn and terminates abruptly at the foot of a 40m wooded cliff.

The Southwick cliffs are a nature reserve of the Scottish Wildlife Trust. Romantic naturalists may delude themselves that the merging of ancient woodland into saltmarsh reflects a landscape unsullied by the touch of man. Both wood and marsh apparently sustain themselves without our intervention; here is a fragment of the pristine.

The structure of any woodland can be read to deduce something of its past. A striking feature of Southwick is the lack of any substantial boundaries; the wood runs directly into the marsh. There are the remains of an old fence-line but nothing suggesting any division

The Southwick Burn.

for management in the historical past. This unbroken transition, combined with a scattering of holly, thorns and grassy glades, suggests a wood pasture, the livestock on the merse being unhindered in finding shelter and forage amongst the trees.

Southwick is a wooded cliff and for the most part naturally inaccessible to cattle and naturalist alike. The safest way to the shore passes through the 'Needle's Eye', a natural rock tunnel. Once on the merse you can see that trees are not wholly dominant and prows of rock are exposed to the sea. Stacks rise out of the merse, the most prominent dubbed 'Lot's Wife', the unfortunate soul here petrified not in salt but in granite.

Mildly radioactive veins of uranium and bismuth run through the cliff as well as more stable elements of arsenic and lead. Skeletal soils weathered from these rocks contribute to the stresses restraining the vigour of the wood, as do the livestock wandering through the merse. There is a distinct historical browse line along the foot of the cliff, below which the smothering qualities of Ivy *Hedera helix* and saplings have been suppressed.

It is incongruous to stand amongst the halophytes of the merse and then reach out to touch a plant of the mountains. It is within the parched sunny soils of the cliff foot that Sticky Catchfly *Silene viscaria* grows. Sticky Catchfly is conspicuous at over 30cm tall and topped by carnation-pink blooms. In Scotland it is usually found at altitudes above 300m, scattered across the Cairngorms and Ochill Hills. Sticky Catchfly was first recorded at Southwick in 1843 by Peter Gray from Dumfries; he named it Red German Catchfly and gave the location as Lot's Wife.

Sticky Catchfly grows on the rocks at Lot's Wife.

In recent years the Sticky Catchfly colony on the Southwick cliffs has declined as shade and coarser vegetation have crowded in on it. The merse is sporadically grazed so manual clearance and reinforcement of the population through plants grown off-site are necessary to stabilise the situation. Exposure to the tides and sea winds alone is insufficient to maintain the Catchfly's open habitats.

In lieu of regular grazing the merse by the Needle's Eye is mown to favour another of the specialities of Southwick, the

Holy-grass *Hierochloe odorata*. This is a robust perennial grass with rhizomes enabling it to dominate where it is well established. Flowering starts at Southwick from April onwards. The blooms are highly distinctive with an open wiggle-waggle panicle reminiscent of much commoner quaking-grasses *Briza* spp. Holy-grass occupies the transition zone of the upper limits of the merse, washed by the tides and flushed by mineral-rich spring water from the cliff base. The soils in which it grows are not the swampy hollows of adjacent willow beds but are free draining and warm.

It is possible that the natural distribution of Holy-grass in Britain has been supplemented by historical introductions. All parts of the grass are highly aromatic, being likened to vanilla, gorse flowers and coconut. There are traditions in central Europe of it being strewn on church floors to scent the air during festivals, particularly the feast days of the Blessed Virgin Mary. The distribution of the grass on Orkney has been associated with Norse churches but whether there is any significant connection has yet to be proven. The Holy-grass populations of the Solway are all associated with similar habitats along the upper merse and are regarded as truly native.

The aromatic content of Holy-grass is similar to that found in Sweet Vernal-grass *Anthoxanthum odoratum* and other components of quality hay. The scent is not an anti-feeding compound; indeed livestock will seek out and preferentially graze such sweet grasses. In Poland Holy-grass is known as Bison Grass for obvious reasons. European Bison *Bison bonasus* are not native to Britain, not in this interglacial anyway.

Cattle are the ecological successors to our native wild fauna and their grazing benefits the geese, Tadpole Shrimps, Natterjack Toads and all the company of the open marsh.

Holy-grass blooms before the adjacent woods come into leaf.

Without them the merse grasslands become unnaturally rank; the special is displaced by the commonplace as the ecosystem loses the fine grain of its definition. Cattle, either wild or domesticated, are an integral part of the saltmarsh ecosystem.

Wigtown

Further to the west the estuary of the Cree flows into the outer Solway. At the mouth of the Cree is the ancient Burgh of Wigtown, the main street and marketplace aligned to the now-silted harbour. Out on the

Monument to the Wigtown martyrs.

merse stands a granite pillar, testimony to a time of terror and intolerance.

In May 1685, on the fall of the tide, two women were secured to stakes driven into the bed of Wigtown harbour. Tradition has it that Margaret McLachan, the older of the two, was placed slightly lower than the younger Margaret Wilson. The fate of the elder may yet encourage the younger to repent. Robert Wodrow (1721) recorded that they had been sentenced to death by drowning, to be tied 'within the floodmark of the sea, and there to stand till the flood o'erflowed them'. Their offence was to refuse to swear an oath declaring the King to be the head of the Church.

The sentence was executed, neither repented, both were drowned. Their graves are tended in the parish churchyard close by merse.

123

Bae Ceredigion

Victorian lovers of picturesque landscapes regarded the estuaries of Bae Ceredigion, Cardigan Bay, with the same admiration as the English Lakes and Scottish Highlands. The mountainous catchments that feed the estuaries from the Dyfi to the Glaslyn belie their southern latitude. There are manifestations of northern climes but there are also floral elements associated with the Mediterranean and the Baltic. The wildlife of the morfa, the saltmarshes, reflects the diversity of their setting whilst being distinctly Welsh. There is nothing immutable about this landscape and its wildlife; its heritage is one of change.

Morfa Harlech

High on a rock above the Glaslyn and Dwyryd estuaries stands Harlech Castle, a masterpiece of medieval military engineering. Harlech had a harbour before the astle was constructed in the closing years of the 13th century. Documentary evidence suggests that at its peak the port was in the same class as Southampton for trade and defence. Standing on the battlements today the estuary lies 4km to the north; the town and castle are landlocked by the plains of Morfa Harlech.

Harlech Castle was commissioned by Edward Longshanks, the English king who succumbed to dysentery on the Solway. King Edward sought to subjugate the princes of Gwynedd and so established a chain of castles to secure his presence deep in their territory. Isolated from supplies and communication, the castle was designed so it could be provisioned by ship through a sea-gate, the fford o'r môr. Today that sea-gate overlooks the railway station and main road, the nearest navigable channel being at Llechollwyn, nearly 5km to the north-east.

OPPOSITE PAGE:
The picturesque
Mawddach estuary.

125

Morfa Harlech.

Morfa Harlech started growing at about the same time as the foundation of the castle. The process may have begun with the development of a cobble spit, as implied by the local place name Cefn mine, thought to be a corruption of Cefn maene, at the back of the stones. If any such spit existed it has been subsequently overwhelmed by the accumulations of sand blown in from the west to create one of the finest and most active dune complexes in Britain. Dunes maintain sheltered waters suited to the growth of saltmarshes. At Harlech the usual process of tidal sedimentation is supplemented by the sands that also feed the dunes. The face of the marsh is growing into the estuary, with Common Saltmarsh-grass *Puccinellia maritima* and Sea-milkwort *Glaux maritima* pioneering the colonisation of the tidal flats. The oldest, now landlocked, saltmarshes continue to be fed by windblown sand and so the Morfa rises above the normal reach of the tide. In the 800 years since the castle was completed the coastal plain has grown to embrace the entire length of the outer estuary from Harlech Point to the tidal gorge at Pont Briwet.

The growth of the dunes and marshes of Morfa Harlech would not have prejudiced the integrity of the castle until they closed the navigable channel to the ffordd o'r môr sometime in the 15th century. On the contrary, having saltmarshes at the foot of the castle would have contributed to the hazards of a besieging army. The risks of

doing battle in such terrain had been considered nearly a millennium earlier and half the world away by the master military strategist Sun Tzŭ. His advice was simple: avoid saltmarshes, and if you have to cross them then do so without any delay (Giles 1910); had King Edward received such counsel he may have lived longer.

Glastraeth at Llechollwyn.

Meeting points

When considered in a European perspective the estuaries in the north of Bae Ceredigion straddle overlapping climatic zones. The relatively southern latitude supports plants associated with the Mediterranean but the high rainfall of maritime mountain country brings with it species associated with the Baltic and Atlantic coasts.

There is a suite of plants of British saltmarshes whose distribution is predominantly of the Atlantic and Mediterranean seaboards of Europe and north Africa. These include Sharp Rush *Juncus acutus*, Golden-samphire *Inula crithmoides* and Sea Wormwood *Artemisia maritima*. The ability of Golden-samphire and Sea Wormwood to grow on cliffs as well as saltmarshes supports respective extensions in their distribution to Luce Bay in Wigtownshire and on the cliffs north of Forvie in Aberdeenshire. As hardy perennials the distribution of such species will adjust as the climate changes but at a sedate pace compared with more adventitious annuals. The Sharp Rush reaches the northern limits of its European distribution in Bae Ceredigion. This is almost certainly its climatic limit; apparently suitable habitats extend northwards far beyond the Solway into southern Scotland. Sharp Rush is an effective colonist within its current range; should the climate change in its favour, a northward expansion into Morecambe Bay and beyond is probable.

In silhouette the Sharp Rush is similar to the other clump-forming rushes of the British countryside but is distinguished by its scale and severity. Sharp Rushes grow in discrete tufts in the higher reaches of sandy saltmarshes; with stems exceeding 1.5m they tower over the marsh turf in a dark phalanx along the reach of the tide. The epithets 'Sharp' and *acutus* are well deserved as each exceptionally hard stem terminates with a tip reputedly tough enough to penetrate flesh and impale wayward golf balls. Botanising around ranks of Sharp Rush requires caution and respect. Grazing animals naturally avoid these painfully inedible spikes to the advantage of more palatable species which grow amongst their tussocks.

ABOVE: Sea Wormwood.

BELOW: Parsley Water-dropwort blooms within the shelter of Sharp Rush.

In contrast to being at the northern limits of an Atlantic-Mediterranean flora, the Bae Ceredigion marshes support a southern outpost of a species of the Baltic region *Bryum marratii*, the Baltic Bryum. This moss shares a similar habitat to Sharp Rush in being associated with the upper edges of saltmarshes. Unlike its imposing neighbour, the Baltic Bryum is easily overlooked as each leaf grows to 2mm or less in low mounds rarely exceeding 10mm. Being small in stature the *Bryum* is at risk of being out-competed for light and space and so requires exceptionally short turf in which to thrive; such conditions are maintained by sheep. The European stronghold of *Bryum marratii* is around the Baltic, a sea with a very different chemistry

from Bae Ceredigion. Whilst connected to the world's oceans, the Baltic has limited exchange with the North Sea as well as an exceptionally high input of freshwater from the neighbouring landmasses. Depending on the season the grasslands where *Bryum marratii* grows on the Estonian coast have been described by Ingerpuu & Sarv (2015) as experiencing salinities between 0.5 and 7ppt (parts per thousand). In comparison, the salinity of Bae Ceredigion ranges from 33–34ppt, albeit with local and seasonal variations. It is the consistency of rainfall and the reliability of freshwater seepages across the saltmarshes that sustain the salinities tolerated by the moss in Wales.

Until recently Bae Ceredigion was thought to be the southern limit of the Baltic Bryum in Britain. Diligent recording has been extending its range and in 2012 a population was discovered on flushed saltmarsh at Whiteford at the western end of the extensive saltmarsh commons of the Gower peninsula. Common rights to graze sheep and ponies are exercised across these saltmarshes, so maintaining the short flushed turf required by the moss. The discovery of a new population of *Bryum marratii* requires cautious interpretation as an indicator of environmental change; the incidence of records of critical species is partially a reflection of the distribution of expert recorders.

Of all the plants of Bae Ceredigion lying on the edge of their known ranges the most intriguing is a flowering plant, 'Lleidlys Cymreig', the Welsh Mudwort *Limosella australis*. The scientific name for this mudwort reflects its discovery in being derived from *limosus* = muddy or growing in muddy places and *australis* = south or southern. The naming of *Limosella australis* was the responsibility of Robert Brown who, as a young botanist, was part of an expedition to Terra Australis. Brown and his companions set sail in 1801 under the command of Captain Flinders to spend four years exploring what were to become the colonies of Australia. They collected nearly 4,000 botanical specimens, of which one of the least spectacular was an anomalous mudwort. Brown returned home in 1805 in HMS *Investigator*, the anxieties of the long journey being compounded by the ship having been condemned as unsound but subsequently refitted. On arriving safely with his collections Brown was lauded across Europe and was rewarded with being appointed as the first Keeper of Botany in the British Museum. It took him five years to describe and catalogue his specimens which included over 1,700 species new to Western science, one of which was his mudwort, duly named *Limosella australis*. Over the course of a long life Brown's achievements in biological science led him to be regarded

Robert Brown.

as perhaps the greatest figure in the whole history of British botany. He was saluted on his death by the acclaimed Von Humboldt as 'Facile Botanicorum princeps, Britanniae gloria et ornamentum' – Simply the finest botanist, Britain's glory and ornament.

As other expeditions carried botanists around the world Brown's mudwort was found to have a wide distribution across the southern hemisphere and along the eastern seaboard of North America. It was not until 1897 that his mudwort was tentatively recorded from Europe, those first specimens being from the Glamorganshire coast. At first the identification of the Glamorganshire mudwort was disputed, with Druce (1932) considering it to be a variety of the closely related Mudwort *Limosella aquatica* under the name of *var. tenuifolia*. In contrast to the newcomer, *Limosella aquatica* had been known as a native British species since 1663 when it was recorded by John Ray on Hounslow Heath in Middlesex. Over time the Glamorganshire plants were accepted as being Brown's *Limosella australis* and given their local name 'Lleidlys Cymreig', the Welsh Mudwort. This judgement was later confirmed by genetic studies, the two species of *Limosella* under debate having distinctly different numbers of chromosomes.

The Glamorganshire populations of Welsh Mudwort were scattered along the coast between Port Talbot and Llanelli but became increasingly rare through the opening decades of the 20th century with the deteriorating quality of saltmarsh habitats. The last record from south Wales was in 1970 and so the species is regarded as locally extinct. Fortunately, in 1921 a population was discovered on the Glaslyn at Traeth Mawr, followed by more sites in the adjacent marshes of the Dwyryd and then further south in the Dysynni. Numbers fluctuate with changing conditions, with apparently short-lived satellite populations occurring in the Morfa Harlech where the Dwyryd and Glaslyn share a common estuary. The saltmarshes of Bae Ceredigion remain the sole locality for Welsh Mudwort in Europe.

There having been over 400 years of botanical recording in Britain, any suggested additions to the native flora tend to be viewed with a degree of suspicion. The 2006 edition of the Red Data Book for British plants declined to assess the status of the Welsh Mudwort as the editors, Chris Cheffings and Lynne Farrell, regarded it as a newly arrived species with an unstable distribution and mostly growing in artificial habitats. The distribution is undoubtedly fluctuating within this highly dynamic sedimentary coast but its habitats are far from artificial.

It is possible that Welsh Mudwort is a long-standing member of our native flora which has been overlooked. With affection it can be described as an unassuming plant growing in a highly specialised niche within an historically neglected habitat. A mature specimen comprises a neat rosette on the surface of estuarine silts; individual plants may be 1cm across with exceptional specimens achieving 8cm. There is no foliage as such; the leaf and its stem combine as a tapering cylinder. Reproduction appears to be mostly vegetative by way of runners, creating a turf which can grow whilst permanently submerged. Flowering does occur in such aquatic stands, the buds remaining closed with the concealed blooms self-pollinating. Should the rosettes have a sufficiently long season in the air then the flowers open into a 4mm scented disc of five pale petals. The plant has been described by Jones (1999) as 'quite conspicuous'; a fair description by the expert with exceptional knowledge of its life cycle and habitats, but somewhat optimistic when applied to the average botanist.

Reliable populations of Welsh Mudwort are to be found in the brackish marshes on the North Wales Wildlife Trust's nature reserve at Traeth Mawr. In 1811 the natural alignment of this transitional habitat was massively disrupted by the construction of the Porthmadog Cob, a

Traeth Mawr.

barrage across the inner estuary of the Glaslyn. The barrage has never been entirely effective in excluding tidal waters and so an artificially extensive zone of weakly brackish habitats has formed upstream of the Cob. Herds of cattle graze the pools in which the mudwort grows, so maintaining a close-cropped sward and preventing Grey Club-rush *Schoenoplectus tabernaemontani* and other coarse vegetation from becoming dominant.

The questionably native Welsh Mudwort shares its open brackish habitat with the unimpeachably British Dwarf Spike-rush *Eleocharis parvula*. This is the smallest of Britain's spike-rushes, which, where abundant, forms a felty nap across estuarine silts. In favourable conditions it can grow up to 8cm, the numerous slender leafless stems arising from a mat of rhizomes; in most locations the plants are very much smaller. Mats of the spike-rush become slightly less inconspicuous in late summer when they bare their petal-less flowers. Unfortunately the Welsh populations are renowned for their shyness, with reproduction being chiefly by vegetative means. As with Welsh Mudwort the Dwarf Spike-rush and its habitat are familiar only to the cognoscenti. The late discovery by botanists of these cryptic diminutive plants is understandable. Dwarf Spike-rush was first recorded in Britain as late as 1835 when it was found growing in a tidal backwater of the Lymington River in Hampshire. To set this record into context, the estuary at Lymington had been recorded by eminent botanists over the previous 200 years. In a far less intensely surveyed coastline it is not surprising that it took until 1921 for Welsh Mudwort and Dwarf Spike-rush to be recorded from the estuaries of Bae Ceredigion.

Until recently Dwarf Spike-rush in Britain was regarded as having its British stronghold in the Bae Ceredigion estuaries, which was complemented by three small populations on the south coast of England. In 2000 the discovery of a significant population in Scotland was announced from the brackish reaches of the Cromarty Firth. In 2016 I discovered a second Scottish site on the Kyle of Tongue. As more botanists are initiated into the micro-habitats of brackish mud one may hope that new populations of Dwarf Spike-rush and possibly even Welsh Mudwort wait to be discovered.

The debate as to the native status of Welsh Mudwort in Britain has its mirror image on the far side of the world. Dwarf Spike-rush has recently been reported by Saintilan (2009) as having been discovered in Australian saltmarshes. The discovery prompted the same discussion; could it be native, has it been overlooked?

A particularly floriferous clump of Dwarf Spike-rush.

Before concluding accounts of the botany of Bae Ceredigion it is timely to speculate on the earliest written botanical records from Wales. Pliny the Elder (d. 79 AD) gave an account of the druids in his *Natural History* and in doing so described two plants. The druidical use of Mistletoe *Viscum album* is a much-revered facet of Welsh culture but has no associations with saltmarshes. The second of Pliny's Welsh plants was 'Samolus'. Quite which species Pliny was referring to is ambiguous; the only clues are that the plant lived in humid places and was used as physic for cattle. The strongest candidate is Brookweed *Samolus valerandi*, whose scientific name derives from a 16th-century tribute to Dourez Valerand, an apothecary from Lyon. Much-repeated tales can accumulate the trappings of truth as centuries later Samuel Gray (1821) confidently asserted that Brookweed was the Samolus of Pliny.

Pliny's description reads:

The druids, also, have given the name of 'Samolus' to a certain plant which grows in humid localities. This too, they say, must be gathered fasting with the left hand, as a preservative against the maladies to which swine and cattle are subject. The person, too, who gathers it must be careful not to look behind him, nor must it be laid anywhere but in the trough from which the cattle drink.

133

Brookweed.

Brookweed is an inconspicuous member of the Primrose family whose association with Britain's coastal wetlands gains strength as one travels north. It is a species of brackish marshes but will tolerate extremes of salinity ranging from completely freshwater wetlands to occasional tidal inundation. It grows in abundance in the margins of Traeth Mawr within a stone's throw of Welsh Mudwort.

Time out of mind

Legend has it that once there was a land called Cantref-y-Gwaelod which lay to the west of Bae Ceredigion. At low tide the five rocky sarns of the Bae have been taken to be remains of an ancient people who lived on the flat lands lost to the sea. The greatest of them all is Sarn Badrig, St Patrick's Causeway, which reaches seaward out from Mochras Island. Sadly there is no masonry there, just geology, but the presence of the sarn reinforces the telling of the tale.

Written versions of the legend of Cantref-y-Gwaelod originate in the oldest surviving book in the Welsh language, the Llyfr Du Caerfyrddin, the Black Book of Carmarthen. This is a compendium of poems dating from the 9th to the early 13th century, one of which recalls the fate of Cantref-y-Gwaelod, which was lost when the heroic drunkard Seithennin neglected its sea defences. Over the following centuries the story has been repeated and decorated but yet may contain a germ of truth, a folk memory of the evolving coastline of Cardigan Bay.

The dates of the writing of the Black Book are unknown but are likely to be contemporary with the chronicler Gerald (*c*.1146–*c*.1223). Colt Hoare (1806) translated Gerald's encounter with a drowned forest off the coast of Pembrokeshire:

being laid bare by the extraordinary violence of a storm, the surface of the earth, which had been covered for many ages, re-appeared, and discovered the trunks of trees cut off, standing in the very sea itself, the strokes of the hatchet appearing as if made only yesterday. The soil was very black, and the wood like ebony. By a wonderful revolution, the road for ships became impassable, and looked, not like a shore, but like a grove cut down, perhaps, at the time of the deluge, or not long after, but certainly in very remote ages, being by degrees consumed and swallowed up by the violence and encroachments of the sea.

It is still possible to wander through such ruined forests on the Welsh coast. At Borth, on the mouth of the Dyfi, an abundance of tree stumps and boles emerges from the peat beds of the eroding foreshore. These organic remains have been dated from about 6,000 years ago. Laminated into the peat are bands of clay from the saltmarshes of an ancient estuary, sediments recalling a matrix of estuarine marshes and woods. Alder *Alnus glutinosa* is preserved in abundance, suggesting elements of a landscape reminiscent of today's brackish woods by Loch Fleet. Over the millennia the composition of saltmarsh, woodland and bog at Borth would have morphed with the shifting sand and shingle of the shore. One exposure of peat contains a wedge of estuarine clays deposited over the course of a thousand years, that particular flood abating around 1,750 years ago. It is conceivable that the oral traditions of Welsh culture carried the memory of those catastrophic events over a handful of centuries before they were laid down in the Black Book.

Relics of Borth's estuarine woodland.

Genesis of the Humber

O n the cliff top at the convergence of the Humber and Trent is a labyrinth, a wandering path inexorably reaching its destination by a circuitous route. The intricate meanderings were carved into the Lincolnshire turf by people whose names and intentions are long forgotten; here is a fitting place to contemplate the prehistory of saltmarshes. A field naturalist is fortunate in not having to grasp the depth of geological time nor the conceptual challenges of astronomy. The character of Britain's wildlife today is the product of a mere score of millennia, a few hundred human generations, a fragment of evolutionary time.

Archaeological evidence suggests that people first lived by the Trent and Humber about 12,500 years ago. These were modern humans, people whose art implies perceptions similar to our own but who differed greatly in the tools available to them. Their view from the cliff over the confluence at Trent Falls would have contained elements recognisable to us today. The distribution of hills and plains remains the same even though the surface of the land has been modified. The rivers still flow in the same direction, draining the same hinterland. Today the Trent is corseted within banks and disciplined to a navigable channel. The early visitors would have looked out over the same river but as a myriad of braids anatomising across the plain of a drying glacial lake. Interspersed amongst these wetlands were drifting dunes being reworked by wind and water; theirs was a landscape settling after millennia under the thrall of ice.

Whilst today it is a short walk down the sunken lane from the labyrinth to the reedy saltmarshes of the Trent Falls, it is possible that

OPPOSITE PAGE:
Julian's Bower, the
Mizmaze at Alkborough.

137

someone standing on that cliff over 12,000 years ago would have had no experience of saltmarshes. Those travellers had penetrated into the heart of a great peninsula far from the nearest ocean.

The evidence that helps us interpret the natural history of prehistory is coarse grained; the valleys of the Trent and Humber are no exception. Archaeologists have located a scatter of objects made by people, mostly stone tools but also the occasional artwork. These human artefacts, combined with relics of the natural world in which they lived, grant fragmentary glimpses of life in recent prehistory. With caution there is sufficient tangible material to interpret the natural history of this period without fanciful extrapolations of imperfect evidence.

People first arrived in the valley at a time when the local environment was supportive of the greenstuff and game on which they relied. The climate of England at that time was as mild as that of today, and possibly even milder. Even so, the favourable conditions that supported the initial colonisation by people was short-lived. After a promising start the weather reverted to a much colder state; it wavered on the margins of human tolerance until a more benign and settled climate was established some 10,000 years ago. The people who resettled the valleys had over the intervening years developed a different technology with a more refined working of their flint tools. They are known to us as the people of the Middle Stone Age, the Mesolithic.

Doggerland

The history of the Humber's saltmarshes starts far away from the modern coastline in a period when sea level was much lower than today. Should our original settlers have followed the Trent downstream, their journey would have taken them for hundreds of kilometres through a vast plain of lakes and great rivers terminating in coastal marshes on the northernmost shores of Europe.

The concept of connectivity between continental Europe and the British Isles has long been recognised. This connection has often been expressed as a land bridge, by etymological inference an adequate but temporary connection across which our ancestors passed to lay the foundations of a nation. In the closing years of the 20th century Bryony Coles (1998) presented an accumulation of evidence describing this connectivity as a series of maps. At the centre of these landscapes were the Dogger Hills from which she coined the name Doggerland.

Doggerland was the most recent of many occurrences when the place we know as the North Sea was dry land and fully occupied by terrestrial life. The formation and breaching of connections between the Eurasian continent and its outlying islands is a process which has happened many times before as the world's water changed from ice to liquid and back again. Over the last 2.5 million years there have been about 20 such fluctuations of varying intensities. Bryony Coles' chronology of Doggerland illustrates the most recent period of connectivity, which persisted for thousands of years in a state habitable for people and wildlife.

In describing the past it is convenient to refer to the geography of the present. At its greatest extent Doggerland's northern coastline stretched from central Scotland to Denmark with a southern shore somewhere off Brittany. As with today this breadth of latitude would have supported a variety of local climates with a diversity of habitats reflecting the influences of geology, tide, soils and drainage. The available data do not enable us to understand the subtlety of such variations and so caution is needed when applying a common interpretation to an historical landscape that was as large as a modern nation.

Bryony Coles' maps have a beautiful simplicity concentrating on the principal components of land, sea and ice. Within these foreign shores are the familiar cartographic lines of the headwaters of the rivers that fed the plains of Doggerland. These rivers remain as consistent features as her sequence of maps charts the shrinking ice fields and the expansion of the sea. As sea levels rose so the river catchments were shortened and their estuaries broadened out into a great bay backed by the Dogger Hills. The concluding map illustrates those hills and their surrounding marshes as part of a complex of islands on the western coast of Eurasia. Britain became separated from the continent when the subtidal components of the coastal marshes merged. This sequence is explained in the broadest of geographical and chronological scales; a thousand years or so is a detail in the course of such events.

Populating Doggerland

It is possible tentatively to populate the blank canvas of Doggerland. As soon as liquid freshwater became available then the transitory species of the ice fields would have been accompanied by more permanent residents. The initial conditions on the northern coast are believed to have been those of an ice desert, freeze-dried for most of the year.

We know from today's polar regions that the characteristic vegetation of maritime ice desert comprises crusts of lichen and mosses whose growth rates are glacially paced. Within this shelter live a fauna composed of mites, springtails and midges; life in such terrain is measured in millimetres. Even before colonisation by flowering plants, the coastline is likely to have been seasonally occupied by the diverse waterbirds and marine mammals of the Arctic oceans. Vegetation is not needed for haul-out sites where seals pup, nor for the nesting grounds of pelagic wanderers.

Saltmarshes are an established feature of the Arctic tundra today; there is no reason to believe they would not have been present in the past. The coastal margins of early Doggerland will have included areas of saltmarsh once the climate was sufficiently warm to support flowering plants. The Doggerland coast differed from the modern Arctic in latitude, ocean currents and tidal ranges. How such differences may have materially affected saltmarsh formation is unknown. The shores of the Arctic today support a diverse range of saltmarsh habitats that offer suggestions as to the character of Doggerland's early coastal wetlands. An analogy may be found in the marshes of the White Sea in western Russia. These marshes include familiar associations of Sea Arrowgrass *Triglochin maritima*, Sea Aster *Aster tripolium* and Sea Plantain *Plantago maritima*, together with extensive brackish communities of Slender Spike-rush *Eleocharis uniglumis* and Sea Club-rush *Bolboscheonus maritimus*, which grade into upper marshes dominated by the sedge *Carex subspathacea*. It is only the last of these species that is not a regular component in British saltmarshes.

The natural history of Doggerland is most readily expressed through the remains of the large animals that once lived there. Each year the haul of jumbled bones trawled from the North Sea runs into the tens of thousands of specimens. As the climate warmed between twelve and ten thousand years ago the land became sufficiently vegetated to support the large grazing animals of tundra and steppe. At the time of our travellers to the Trent

Reindeer on Arctic saltmarsh, Svalbard.

Falls the fauna of Doggerland included Reindeer *Rangifer tarandus* in the tundra, together with Mammoth *Mammuthus primigenius*, Saiga Antelope *Saiga tatarica*, and Wild Horses *Equus ferus* on the steppe. All of these creatures would have been attracted to the summer pastures of the coast where tides helped clear the snow, and oceanic influences promoted the early growth of vegetation.

As the glaciers melted so sea levels rose and the northern shoreline migrated southwards. The processes of coastal realignment would have been very similar to those that occur along our softer coasts today. Sediments are worked and reworked by storm, tide and the life of the marsh. The pace was such that any individual generation of animal was unlikely to have experienced hugely disruptive changes. Indeed, for the species of saltmarshes the migration of the coast is likely to have increased the potential for intertidal habitats and the richly productive ecosystems they support.

Getting warmer

About 10,000 years ago the environment of Britain underwent an abrupt change. Within a very short period, measured in decades rather than centuries, the climate underwent rapid warming and then stabilised. Derek Yalden (1999) likened this change as being equivalent to the difference between the Siberian Taimyr peninsula and London. Mean July temperatures rose from 8°C to 17°C with dramatic consequences for all livings things; everyone, everything, needed to adapt, migrate or die.

An imaginative reassessment by archaeologists of geological surveys has provided an insight into the landscapes of Doggerland at this stage of its development. Vincent Gaffney *et al.* (2007) describe how the distinctive formations of rivers, estuaries and saltmarshes, long submerged and buried beneath the seabed, are detectable through seismic reflection surveys. The surveys reveal prehistoric landscapes on a grand scale by virtually 'peeling back' the sediments of the seabed, layer by layer. An analysis of 2.3 million hectares centred on the Dogger Hills has mapped the shape of the land following the dramatic period of warming. At that time the north coast of Doggerland was a bay whose borders extended to 690km; that bay was fed by ten major estuaries, each supporting extensive saltmarsh systems. The scale of such a landscape is difficult to comprehend as it dwarfs any equivalent wetland system in modern

Europe. To set this in a modern context the shore of the bay of Doggerland was ten times greater than that of the Wash.

The newly established climate turned out to be remarkably consistent compared with the rapid changes of preceding millennia. The wildlife that became established would have been familiar to a modern naturalist with the potential of the landscape to support the predecessors of the principal habitats of modern lowland Europe.

To what degree the large mammals of this age affected the landscape is much debated. Ecological theory developed in the early 20th century hypothesised habitats dominated by a continuous canopy of trees. Extensive high forest was proposed as restricting the populations of herbivores and all other species dependent on sunny places. Later in the century this model was challenged by ecologists including Frans Vera (2000) who argued that the presence of large grazing animals would have modified the density of tree cover resulting in a complex of wooded and open habitats; such a landscape would support a far wider range of habitats with the diverse species assemblages of gladed woods and open countryside. The Vera model helps explain the preponderance of species of open habitats and dappled shade in Britain's native fauna and flora. Saltmarshes have played their role in this debate as a habitat naturally resistant to tree growth, together with being positively attractive to large grazing animals. Irrespective of how densely the hinterland was wooded, we can be confident the tidal margins experienced an unbroken continuity of open landscapes.

Remains from the seabed and around the coastal fringes of Doggerland testify to the diversity of the birds and mammals that occupied the newly temperate landscapes. Some of these animals were already present at the end of the era of ice and included some of the largest grazing animals of post-glacial Europe. These were Aurochs *Bos primigenius*, the now extinct native Wild Cattle, together with Elk *Alces alces*, Red Deer *Cervus elaphus* and Wild Boar *Sus scro*. As the climate changed in their favour other animals colonised from southern refugia including Roe Deer *Capreolus capreolus*, Beaver *Castor fiber* and Otter *Lutra lutra*. Similarly, the larger predators of the earlier era, both Brown Bear *Ursus arctos* and Wolves *Canis lupus*, maintained their populations through this period of change. As Roe Deer spread into the new landscape so did Lynx *Lynx lynx* to which the deer are a principal prey. Wild Horses may have made the transition from steppe to temperate grasslands but the evidence is too fragmentary to have reasonable confidence either way. There is however enough by way of preserved footprints and other indirect evidence not to dismiss their contribution to the combined effects of native grazing animals on the developing landscape.

Skeletal fragments of large mammals have a greater likelihood of preservation than the slighter bones of birds. The remains of birds that have been recovered from this period are mostly of species dependent on open wetland habitats to complete critical stages in their life cycle. Not all remains have been identified to species level; archaeological accounts tend to group these bones as ducks, geese and swans. Local detail can be added where the quality of remains permit. The ducks included Mallard *Anas platyrhynchos*, Eurasian Wigeon *Anas penelope* and Shelduck *Tadorna tadorna*. Greylag Geese *Anser anser* and Whooper Swan *Cygnus cygnus* have been recorded, as have both White *Ciconia ciconia* and Black Storks *Ciconia nigra* along with Common Cranes *Grus grus*. Such a species list today would be associated with a coastal wetland of great size and structural diversity.

The earliest definitive proof of human occupation of Doggerland was cut from a lump of peat by Captain Lockwood of the trawler SS *Colinda*. In September 1931 Lockwood was fishing out of Lowestoft between the Leman and Ower Banks, some 40km off the North Norfolk Coast. The trawl was in waters some 20m deep. With the advent of steam and latterly diesel it became possible to work the seabed with a vigour denied to sailing craft. An occupational hazard of mechanised fishing was the bycatch of blocks of peat, known to the fishermen as moorlog. In breaking up a particularly inconvenient slab of moorlog Captain Lockwood's spade struck an unexpectedly hard object, an antler elaborately worked into the point of a fishing spear. The spear was subsequently dated using carbon 14 assay from 11,740 years ago, give or take a margin of 150 years. Whilst the discovery of the spear was serendipitous there is no reason to believe people had not been present beforehand and did not continue to live in Doggerland until the events that finally submerged its marshes and hills thousands of years later.

There is a school of thought promoted by Jim Leary (2015) and others that those coastal marshes were particularly favoured by Mesolithic people. Estuarine wetlands are amongst the most productive ecosystems in the world in generating a diverse and secure food supply for humans. Excavations from around the North Sea illustrate people exploiting the abundant sources of protein ranging from fish, waterbirds, shellfish and mammals. The vegetation of these habitats is similarly productive in the roughage, carbohydrates and vitamins of a balanced diet. Many plants that were eventually domesticated are natives of the naturally nutrient-rich habitats of coastal margins.

The importance of the Doggerland estuaries to the people of the Mesolithic has been the subject of much conjecture. Some archaeologists have gone so far as to place Doggerland at the heart of Mesolithic society in western Europe and to propose the eventual loss of that land as a cultural calamity. What is clear is that as the sea level rose and progressively flooded Doggerland so the ratio of estuary to land increased. The suitability of the Doggerland marshes for human habitation is likely to have been increasingly favourable right up to the time of its final inundation.

Doggerland submerged

Throughout the Mesolithic period sea levels continued to rise. River catchments were shortened and estuaries migrated upstream. Sea-level rise appears to have accelerated between 8,900 and 7,800 years ago, a period that included at least two catastrophic events. About 8,200 years ago a submarine landslide off the Norwegian coast generated a tsunami that radiated through the Arctic Ocean and the shallow seas of the Doggerland coast. Whilst undoubtedly calamitous in the short term, the floods resulting from a tsunami need not have had long-lasting effects. Tidal waves do not permanently raise sea levels although they may accelerate coastal processes already in progress such as breaching natural barriers and establishing new patterns of tidal flows. In the same period as the tsunami there was the collapse of part of the north American ice sheet, releasing the contents of vast glacial lakes into the Atlantic. The impacts of this event went beyond a step change in sea levels to include the modification of ocean currents and the temporary cooling of the climate of the northern hemisphere.

At some stage around this period the north and south coasts of Doggerland became connected and Britain was separated from mainland Europe, at least at high tide. The process of separation may have been catastrophic but is equally likely to have been incremental. Throughout this period the estuaries of each of the major rivers of Doggerland would have continued to migrate towards their headwaters. Should those catchments arise in the Dogger Hills then the estuary would eventually be extinguished. The estuarine marshes and hills of Doggerland would have become a progressively fragmented marsh-bound archipelago where the inundation and erosion of sediments outpaced the processes of accretion. The final inundation of Doggerland is open to debate; different authorities propose dates ranging between 8,500 and 5,800 years ago. Such events had happened many times before and were unlikely to have been of particular significance to most species present at the time. The life of coastal marshes is adept at adapting to change.

The migrating coastline

At the time that Doggerland was becoming fragmented so the development of the estuary of the Humber comes into clearer focus. The archaeological record has captured the sequence of events that carries the estuaries and saltmarshes of Doggerland upstream into the recognisable geography of modern England.

Silts deposited in estuaries and bound up within saltmarshes are highly distinctive, as are the peat bed remains of freshwater and brackish vegetation. Laminations of peat and silts in the bed of the Humber record the succession of estuarine communities. Within the long history of rising sea levels there are numerous periods of acceleration and reversal. The technical vocabulary used to describe such phases in the life of an estuary borders on the judgemental. When the intertidal advances over dry land it is described as transgressive. When the intertidal is displaced by freshwater or dry land it is regressive. At any one time in an estuary both processes may occur simultaneously. It is not only global sea-level rise and isostatic readjustment that drive changes to the vegetation and soils. A shifting sandbar or a wandering meander may favour one habitat over another within the same tidal regime. Each change is captured in the accumulation of sediments, which, if not overwritten and eroded by future events, embodies the history of the estuary.

Robert Van de Noort (2004) has shown how an estuary with its genesis in Doggerland rolled upstream into the broad valley of the Humber. The story starts about 8,000 years ago when saltmarshes occupied what is now the mouth of the estuary in the vicinity of today's Immingham docks. Other deposits from this period suggest that peat-forming estuarine marshes were present as far upstream as the city of Hull. The next thousand years saw a rapid advance of the saltmarsh upstream into the Humber valley taking it beyond the modern Humber Bridge to Ferriby, some 40km upstream of

Seven thousand years ago saltmarshes migrated from Saltend (top), past the Humber Bridge (centre) and on to Ferriby (bottom).

Immingham. Tidally influenced estuarine peatlands made similar progress upstream, with deposits from this period laid down in the Humberhead Levels and in the lower reaches of the Ancholme valley.

By 6,000 years ago conditions changed and the pace of upstream migration halved to about 20m a year. By then saltmarshes had overwhelmed the estuarine peats at Hull and the mouth of the Ancholme valley. The following thousand years saw the pace of change slowing even further to about 10m a year, but by then saltmarshes had penetrated deep into the floodplains of the Ouse and Trent in the Humberhead Levels as well as into the Ancholme valley. It was about 4,000 years ago when this trend experienced a brief reversal with freshwater habitats advancing at the expense of the saltmarsh; during this period the headwaters of the estuary in the Ancholme valley and Humberhead Levels experienced the growth of peat over saltmarsh. This reversal was however relatively short-lived, with intertidal habitats reviving and the Trent becoming tidal as far upstream as Scunthorpe and beyond. It is from this period that saltmarshes dominated the confluence of the Humber and the Trent and established the landscapes that would survive until drainage engineers arrived just a handful of centuries ago.

People were living alongside the Humber throughout the 5,000 years it took the estuary to migrate to the Trent Falls. Over those years these people changed as they embraced agriculture, brought land into cultivation and displaced wild animals with domesticated livestock. Communities became more settled, their work specialised with goods being traded over increasingly long distances. Over this period technological developments supplemented tools of wood and stone with progressively

The fens and saltmarshes of the Ancholme have been comprehensively drained.

hard metals. By the time saltmarshes and raised bogs were established across the Humberhead Levels the local people were benefiting from settled agriculture and a sophisticated use of bronze.

The Bronze Age people of the Humber left archaeologists a rich legacy of remains including a boat excavated from the estuarine silts of the Ancholme River near the Lincolnshire town of Brigg. The craft is a flat-bottomed vessel built of planks sewn together with Yew *Taxus baccata* withies and caulked with moss. People at the time were skilled in working with metal but had yet to start using it to make nails for boatbuilding. The dimensions of the partial remains suggest a boat with the capacity to transport loads over 7 tonnes, equalling 10 people and 30 cattle.

For some reason the boat was abandoned and laid up on the edge of the marsh. Upon excavation 3,000 years later, fragments of stems and rhizomes of Small Cord-grass *Spartina maritima* were recovered from beneath its planks. Hillman (1981) describes how a mass of seed from fen and saltmarsh communities was removed from the surrounding sediments. Despite the overlying complications of freshwater flotsam the flora found beneath the boat is comfortably familiar, with the presence of species associated with the middle and upper reaches of an English saltmarsh including Sea Arrowgrass, Sea-milkwort *Glaux maritima* and Sea Club-rush. Amongst the preserved remains are seeds of Eight-stamened Waterwort *Elatine hydropiper*, suggesting that somewhere nearby there were seasonally exposed mudbanks. In Britain today the waterwort is a great rarity limited to freshwater wetlands but on the continent its range extends to brackish conditions. At the time the Brigg boat settled into the Humber wetlands both habitats were present in abundance.

Our saltmarshes and their wildlife are detectable in prehistory; they are this interglacial's representation of a habitat whose origins lie much deeper in evolutionary time. Saltmarshes are complex habitats developed in the presence of large grazing animals, initially herds of wild herbivores but more recently their domesticated ecological successors. Saltmarshes have never been static; they move across the landscape, keeping pace with global processes. Since the end of the last glaciations, emotionless and impassive, twice daily the tide rolls Britain's saltmarshes inexorably upstream.

Eboracum

Prehistory is generally regarded to have concluded with the Roman occupation of Britain for the simple reason that the Romans wrote things down. Roman accounts of Britain are rare, fragmented and exhibit the partial outlook to be expected from a conquering administration. The Romans left no descriptions of the natural history of their garrison and trading centre at Eboracum, the city we know as York. Eboracum was built along the banks of the Ouse and Fosse, both tributaries of the Humber. Today the tidal reaches of the Ouse are constrained by weirs, with the entire catchment of its estuary modified by centuries of engineering.

The redevelopment of York in the late 20th century was accompanied by detailed archaeological investigations led by Richard Hall (1980 onwards). For the first time in nearly 2,000 years it became possible to see the land upon which the city was built. Amongst the botanical detritus of a busy Roman settlement the archaeologists uncovered the remains of a saltmarsh. The preserved plant fragments include Sea Aster, Sea Arrowgrass and Saltmarsh Rush *Juncus gerardii*. Less common are Strawberry Clover *Trifolium fragiferum*, Common Spike-rush *Eleocharis palustris* and Parsley Water-dropwort *Oenanthe lachenalii*, all of which are consistent with the very upper reaches of a tidal river.

Most remarkable are the remains of Marsh Sow-thistle *Sonchus palustris*, which is restricted in Britain today to the coastal marshes of the south and east of England. Sow-thistles are a family of

York city centre remains vulnerable to flooding, having been built next to a tidal river.

yellow-flowered dandelion-like herbs. What is striking about Marsh Sow-thistles is their size; a well-grown plant may exceed 2.5m. As a long-lived herbaceous perennial the sow-thistle grows each spring from buds buried within the tidal debris of the upper marsh. Whilst robust in themselves, the statuesque stems are generally found emerging through other tall vegetation, often tidal reeds or the open scrub of the saline edge. Where conditions allow, their seeds readily germinate and the plant rapidly colonises new ground. The presence of the sow-thistle in Roman York may not be too much of a surprise given that we know conditions are still suitable for it nearby. There are two populations currently growing at the head of the Humber, both of which have been deemed inadvertent introductions from East Anglia. This may be so or perhaps the plant re-established itself naturally. What no documents describe, but the archaeological record reveals, is that 2,000 years ago Marsh Sow-thistle was growing in York at the tidal limits of the Humberhead wetlands.

Marsh Sow-thistle.

In 1906 J G Baker, who had recently retired from being keeper of the herbarium at Kew, published a record for Bulbous Foxtail *Alopecurus bulbosus* from 'Clifton Ings, near York in the meadow opposite the lunatic asylum' and commented 'in Britain usually a plant of saltmarshes'. There is no subsequent record for this or any other saltmarsh species from York. Engineering has relegated saltmarsh from the city whilst time continues to act to the contrary.

1607.

A true report of certaine wonderfull ouerflowing
of Waters, now lately in Summerset-shire, Norfolke, and other
places of England: destroying many thousands of men, women,
and children, ouerthrowing and bearing downe
whole townes and villages, and drowning
infinite numbers of sheepe and
other Cattle.

Printed at London by W. I. for Edward White and are to be solde
at the signe of the Gunne at the North doore of Paules.

Seawalls and the Severn

Most saltmarshes in the south of Britain no longer occupy the full extent of their natural tidal range. From the Humber to the Taff the long history of embanking saltmarshes has generated landscapes of levels, flat farmlands typified by orderly lines and the methodical management of water. Modern seawalls help to exclude tidal flooding from over 35,000ha of land along the Severn. The Somerset and Gwent Levels account for the majority of this former saltmarsh, with further encroachments on tidal lands as far upstream as Gloucester. These combined flood defences represent the accumulation of at least 2,000 years of endeavour.

Saltmarshes and rudimentary sea defences

On the morning of 30 January 1607 the sea walls protecting the levels of the Severn failed. An exceptionally high tide driven by westerly gales swept through Somerset inland to Glastonbury and over the Welsh shore from Redwick to Peterstone Wentlooge. Villages along both banks of the estuary were flooded to a depth of 2–3m. Horsburg & Horritt (2006) estimated the human death toll at 2,000 souls.

Contemporary anonymous accounts (1607) tell of 'violent swellings of the seas, and such forcible breaches made into the firme-land' leading to 'huge and mighty hills of water tumbling over one another' with 'whole fruitful valleys, being now everwhelmed and drowned with these most unseasonal and unfortunate saltwaters . ... Such are the Judgements of the Almighty God ... Good God deliver us all'.

In contrast to the comprehensive sea defences of today, some of the earlier works to moderate the tides were piecemeal and relatively inexpensive. A simple line of defence, no larger than a hedge bank,

OPPOSITE PAGE:
The Great Flood of 1607.

151

is sufficient to modify the highest elevations of a saltmarsh that are only flooded at the peak of spring tides. The village of Puxton lies on the Somerset Levels a few kilometres inland from Weston-super-Mare. At its centre is the Church of the Holy Saviour whose tower leans disturbingly due to the instability of the underlying land. Holy Saviour's is best viewed from the footpath crossing the adjacent Church Field. This field is distinct from the angular divisions of its neighbours as its hedges define an oval. The curving boundary of Church Field follows a bank, interpreted by Stephen Rippon (2006) as an ancient flood defence built on the surface of a high intertidal saltmarsh sometime before the 12th century.

The initial impact of such simple flood defences on the saltmarshes of medieval Somerset will have been marginal. Along the Severn there is a history of constructing 'summer dyke' earthworks sufficient to protect standing crops but ineffective against winter floods. Summer dykes are of a scale and technology achievable within the resources of a small community, and the shape of the banks is telling as a circular earthwork maximises the area defended for the length of wall constructed. In isolation the Puxton defences would have reduced the number of flood events but would not have excluded them entirely and so the saltmarsh would have persisted in a modified form.

The Church of the Holy Saviour, Puxton.

Saltmarshes along the Severn are locally called variants of Warth and Dumble. Dumbles are confined to the Gloucestershire reaches but Warth is a common place name along both the English and Welsh shores. These marshes are characterised by sand and clays, there being little opportunity in the fierce currents and exceptional tidal range for organic-rich silts to accumulate. The marshes therefore develop as terraces dominated by fine grasses, typically Creeping Bent *Agrostis stolonifera*, Red Fescue *Festuca rubra* and Common Saltmarsh-grass *Puccinellia maritima*, together with the Saltmarsh Rush *Juncus gerardii*. Flowery plants such as Thrift *Armeria maritima*, sea-lavenders *Limonium* spp. and Sea Aster *Aster tripolium* are conspicuous by their rarity, as are Sea-purslane *Atriplex portulacoides* and glassworts *Salicornia* spp. The low-growing Strawberry Clover *Trifolium fragiferum* and English Scurvygrass *Cochlearia anglica* can be locally abundant but more exacting species such as the Slender Hare's-ear *Bupleurum tenuissimum* and Parsley Water-dropwort *Oenanthe lachenalii* are confined to the better-drained steps of the terraces. To enjoy the saltmarsh flora absent from the inner estuary it is necessary to travel downstream to the mouths of the Axe and Parrett and on into the more sheltered waters of Bridgwater Bay.

Aylburton Warth, Gloucestershire.

A uniform aftermath on a saltmarsh hay meadow, Gloucestershire.

These fine-grass saltmarshes have the potential for being managed as hay meadows as well as grazing land. Before the 19th century much of the Severn's foreshore was grazed in common as part of the manorial system of the adjacent villages. Most, but not all, of those commons have been enclosed and many continue to be managed by individual farmers rather than communities of commoners. Saltmarsh hay is still cut from the English shore of the Forest of Dean where the crop does not appear to significantly alter the composition of the sward.

Haymaking in floodplains carries an element of risk to counter the advantages of floods bringing fertility. Inundation in midsummer can flatten the crop and spoil it by coating the leaves with the sediments that were welcomed earlier in the season. A late summer flood can sweep the whole year's investment downstream. The floodplains of the lower Severn Valley and its estuary were an important area for haymaking where surpluses were exported into neighbouring regions until the advent of artificial fertilisers and silage technology.

It is in the grassy marshes of the Severn that the Bulbous Foxtail *Alopecurus bulbosus* has its British stronghold. Unlike its close relatives amongst the foxtail grasses, this species is fully tolerant of saltwater. As a perennial it can form extensive patches particularly where other plants are suppressed by grazing such as on the Dumbles at

Slimbridge. In the field it is readily recognised by its pious and thrifty character. The low tufts are greyish-green; radiating from a central point, the erect stems initially rise by way of bended knees. Having knelt oneself, a gentle squeeze to the base of the stems will reveal small swollen apple- and pear-shaped structures, the *bulbosus* storage organs of its scientific name. Intermediate characteristics may occur as the Bulbous Foxtail hybridises with another kneeling foxtail, Marsh Foxtail *Alopecurus geniculatus*, to form the sterile *Alopercurus* × *plettkei*. The hybrid is relatively tolerant of freshwater and persists longer than its saltmarsh parent should the saline influences on a marsh decline. Bulbous Foxtail is restricted in Britain to a line south of the Humber to the Gower. On the European mainland it is widely distributed from the mouth of the Elbe to the eastern shores of the Adriatic.

Bulbous Foxtail takes its name from the swollen structures at its base (above) and grows on tufts (below).

Behind the seawalls

The Church Field flood defences at Puxton are interesting in that they genuinely represent a reclamation, a much misapplied word in the history of sea defences. Whether they knew it or not, the medieval builders of the bank were recovering land that had been farmed centuries earlier. Beneath the adjacent Puxton Dolmores is an ancient landscape of saltmarsh creeks and fields dating from the late Roman period.

On the opposite shore from Puxton the scale of Roman engineering along the Severn is detectable in the marshes of the Gwent Levels. In 1878 a crudely carved slab of the local lias was recovered from intertidal mud off Goldcliff, near Newport. The text on that slab records the completion of 33½ paces of a linear structure by the Century of Statorius Maximus of the First Cohort of the Legion II Augusta. The completion of one third of one hundred units strongly implies the Century were marking their contribution to a larger shared venture.

Re-engineering seawalls over two millennia has obscured the full extent of the embanked marshes during the period of Roman occupation. An abundance of horse bones excavated from the military settlement at Rumney Great Wharf has been interpreted as the legions securing land for their cavalry. At Peterstone Wentlooge the long, narrow trapezoidal fields are reminiscent of the organised manner in which Romans laid out fields elsewhere. From the deep sediments overlaying some of these remains it appears that in the 4th century AD, following the recall of the legions to Rome, sections of engineering failed and saltmarshes were re-established along parts of the coastal plain. It was not until the high medieval period that there was a return to reconstructing those defences. The imposition of the English manorial and ecclesiastical system on south Wales brought with it the capital and organisational structures to undertake such great works.

Not all Roman coast defences were on the same scale as the Gwent Levels. At the head of the estuary at Elmore, just downstream of Gloucester, a saltmarsh meander was embanked through the construction of a 'Great Wall'. That Roman wall survives as a standing monument in the form of a broad bank, about a metre high with occasional remnants of the original stone facing. What was built as a defence against the tide now faces into farmland as subsequent phases of embankment have left it landlocked. If you walk the paths around Elmore there is not the slightest suggestion in the landscape that this was once intertidal. As recently as 1990 a description was published by Allen & Fulford of ancient pastures with relics of former creeks dominating the hinterland behind the seawalls; the efficiencies of modern farming have replaced those marshes with ploughland and ley.

The intrinsic character of a Severn saltmarsh is simple and grassy. Equally simple are the grasslands that develop behind the protection of the seawall. Red Fescue and Creeping Bent persist in the sward with the other saltmarsh grasses being replaced with Crested Dog's-tail *Cynosurus cristatus* and Sweet Vernal-grass *Anthoxanthum odoratum*. There is a considerable increase in the proportion of flowery species including Common Knapweed *Centaurea nigra*, Bird's-foot-trefoils

Lotus corniculatus and *L. pedunculatus* together with the common clovers *Trifolium pratense* and *T. repens*. Seawalls may help protect these grasslands from tidal flooding but without further engineering they are naturally wet for most of the year. The land behind the seawall was therefore traditionally suited to summer pastures for fattening livestock together with haymaking for feed through the lean winter months. Ecologists use the shorthand 'coastal grazing marshes' to describe such landscapes.

The effectiveness of sea defences determines the salinity of the protected land. On the Somerset Levels saltwater was influencing vegetation far inland until as recently as the mid-19th century. At Athelney, some 18km from the open coast and 10km from the tidal Parrett, there are the Higher and Lower Salt Moors. Richard Murray (1896) recognised that the presence of Sea Clover *Trifolium squamosum* indicated an ancient shoreline by the Salt Moors in addition to contemporary records of Sea Aster *Aster tripolium* from the vicinity. In the early 20th century E S Marshall (1914) recorded saltmarsh plants including Distant Sedge *Carex distans* and Divided Sedge *Carex divisa* from as far inland as Shapwick Heath. By the late 20th

The tidal Severn at Elmore.

century improvements in the efficiency of excluding saltwater had either exterminated saltmarsh species or reduced them to rarities on the margins of the tidal rivers. By 1981 Captain Roe, the author of *The Flora of Somerset*, reported that the levels of estuarine alluvium were consistently floristically dull.

Captain Roe's dismissal of the grasslands of the Somerset Levels is a little harsh to modern sensibilities. He was quite correct that these habitats are not associated with any particular botanical rarities or communities of outstanding species richness. The coastal grazing marshes of the Severn have not been immune to the changes of grassland management in Britain through the 20th century. Conventional agricultural grasslands are the product of regimes relying on artificial fertilisers and herbicides; they are, in agricultural terms, 'improved'. With advances in machinery and drainage many landscapes of permanent pastures have been brought under the plough for arable cropping or rotational reseeding of grass leys. Any wildflower-rich grassland, no matter how simple, is worthy of regard.

Pumping out Lower Salt Moor, January 2013.

Returning to Puxton, the Dolemoors, adjacent to Church Field, are a nature reserve of the Avon Wildlife Trust. Within these pastures are the silted-up outlines of former saltmarsh creeks which pre-date Roman settlement. The fields and salt-works established here by the Romans were eventually abandoned, with the land reverting to mudflat and saltmarsh. Sometime around the 10th century the process of excluding the tide started for a second time. The incremental works, including the Church Field summer dyke, eventually coalesced into the coherent defences of today.

The Dolemoors are a relic of a once extensive common established during the medieval period. It is probable that the common started life as a source of saltmarsh hay and fattening pastures. Over time, as the region's sea defences were enhanced, the saline influence would have diminished and so the Dolemoors became freshwater wetlands.

Sea Clover is a species of the extreme upper reaches of tidal waters.

The management of the Dolemoors was determined by an annual cycle of traditions. Every year, just before midsummer's day, the standing hay was divided into portions, each plot carefully measured with a chain kept in the church. These portions were then distributed amongst those owning rights to the hay. The allocation was achieved through lots in the form of 24 apples being drawn out of a bag. Each apple was marked with abstract symbols representing everyday items such as 'Pole-Axe', 'Duck's nest' and 'Oven'. Having been assigned a plot, the right-holder would have until the first of August to mow their hay. If the crop was not taken in time then it would be trampled by the neighbours' cattle as from August to February the land was thrown open for all to pasture their animals. After February the livestock were excluded and each plot-holder then had the responsibility of preparing their portion for the growth and reallocation of the next season's hay.

This elaborate procedure helped in the fair distribution of a valuable common resource. Not all the plots were doled out by lot so the remaining hay was sold to cover the costs of administration. The auction was undertaken in silence with bids tabled in cash until a candle stub had burnt away. Business being completed, the day concluded with revels. Antiquarian accounts range from a polite reflection on displays of hearty mirth to the disapproval of rude rural festivals fuelled by free beer and concluding with the sublime art of pugilism. These traditions persisted until the common was enclosed by Act of Parliament in 1811. Walking the fields today, there are only the faint sulsations of the infilled creeks to remind us that all this land lies below the tide.

Insect life on the Levels

Turning back to the Welsh side of the estuary, there are an estimated 1,400km of ditches and 200km of main drains crossing the Gwent Levels. The collective watercourses are exceptionally rich in wildlife with the majority of the Levels recognised as of national importance for aquatic plants and invertebrates.

The botanical composition of the watercourses of coastal grazing marshes rarely contains distinctly brackish elements. Where these do occur they tend to be associated with dykes close to the seawall or around tidal outfalls. A greater degree of salinity is required to support botanical distinctiveness than is required by invertebrates. The invertebrate communities of the watercourses of the Gwent Levels contain an important element associated with dilute saline conditions. Many of the most specialised creatures of coastal grazing marshes are found at the convergence of brackish and freshwater influences. The water chemistry here is complex, with the marginal effects of salinity in a state of continuous flux. Garth Foster's (2000) review of the beetle fauna of saltmarshes described the ecotone associated with freshwater ponds and ditches occasionally affected by spring tides as being the most species-rich area in Britain. This richness arises from the presence of the specialised species of coastal conditions amongst an exceptional diversity of generalists. Such zones are naturally present in wetlands at the upper limits of the tide; the engineering that creates coastal grazing marshes inadvertently redistributes, and potentially extends, these natural transitions.

OPPOSITE PAGE TOP:
Borrow dykes behind the seawall can be distinctly brackish.

OPPOSITE PAGE BOTTOM:
Freshwater grazing marsh at Magor.

There is a suite of aquatic beetles of coastal marshes that the Gwent Levels shares with the Somerset Levels, including the predatory *Ditiscus circumflexus* and the scavenging *Limnoxenus niger*. For sheer scale there is nothing to compete with the Great Silver Water Beetle *Hydrophilus piceus*. At between 4–5cm long this is Britain's largest insect, its body

TOP: The Great Silver Water Beetle, Britain's largest insect.

ABOVE: Flecked General soldierfly.

length exceeding that of the Stag Beetle *Lucanus cervus*. Finding an adult *Hydrophilus* is not particularly helpful in identifying its natal habitat as they are strong fliers, their stamina demonstrated by a specimen taken from an offshore oil platform. The larvae are fully aquatic and can grow to 7cm long on a rich diet of aquatic molluscs. The bite marks left on snail shells are sufficiently distinctive to confirm the presence of the beetle even if larvae cannot be found.

Amongst the most handsome creatures of the saltmarsh is the Flecked General soldierfly *Stratiomys singularior*. The adult is conspicuous for its size and the uniformly smart yellow flashes along the edge of its thorax. The Flecked General is closely associated with coastal grazing marshes as its larvae develop in weakly brackish mud, typically found where cattle have access to the margins of drains.

Less conspicuous, but equally exacting in its habitat requirements, is the Meniscus Midge *Dixella attica*. This non-biting midge is distinguished by living through its larval stages in brackish ditches suspended in the surface tension of the water. The adults spend a short sedentary life on land and are known chiefly for their patience, for hours on end semaphoring their long limbs to potential mates.

These beetles, soldierfly and midge are representatives of the invertebrate communities of coastal grazing marshes. Their survival is dependent on the maintenance of clean water and a favourable range of salinities together with structurally diverse ditches and aquatic vegetation. An ideal grazing marsh is a broad open landscape of very few trees with the boundaries between fields defined by water. Livestock need access to the edges of the ditches as here their grazing and trampling generate a patchwork of muddy margins and shallow berms supporting both open water and emergent aquatic plants. If ditches of varying width are present then there will be stretches with tall perennial vegetation of Common Reed *Phragmites australis* and club-rushes *Bolboscheonus maritimus* and *Schoenoplectus tabernaemontani*, the core of these stands surviving intact with the margins chamfered by livestock. The stocking levels will be such that the marsh grasslands will be grazed down by the end of the growing season, but until then there is variety in heights of grasses with flowery patches and tussocks of taller vegetation. Such management is increasingly rare where the livestock sector is in decline.

The most celebrated insect of the Gwent Levels has a vital, but indirect, association with the saltmarsh origins of the landscape. In late summer a distinguished salt-and-pepper grey bumblebee can be seen throughout the Levels; this is the Shrill Carder Bee *Bombus sylvarum*. These bees were once widely distributed across southern Britain but their population has gone into decline, contracting to a few core areas including the Levels of the Severn Estuary.

On first impression the requirements of the bee don't appear particularly demanding. Its nests are made amongst coarse vegetation in sheltered hollows such as abandoned rodent holes. In April the queen emerges from hibernation and raises generations of workers and males that forage for pollen and nectar. As a bee with a relatively long tongue the Shrill Carder seeks out flowers with nectaries set at the base of long tubes. The extensive list of forage plants includes clovers, thistles *Cirsium* spp., ragworts *Senecio* spp., Fleabane *Pulicaria dysenterica* and Common Knapweed as well as garden plants such as Sunflowers *Helianthus annuus*. As summer progresses males and young queens emerge to mate and the flight season continues until September when the young queens build nests of their own and settle into hibernation.

With such an undemanding life the Shrill Carder Bee cannot be regarded as a species whose vulnerability derives from exacting habitat requirements. Extensive field work over recent years has concluded that the bees live in metapopulations that are self-sustaining when extinctions and colonisations occur at broadly the same pace. The current best judgement is that a viable population

A queen Shrill Carder Bee.

of Shrill Carder Bees requires at least 10km² of suitable habitat. Inevitably, at any one time most of such an extensive landscape is unlikely to be suitable for foraging; however, there will be some sites where management will always be sympathetic. In the Gwent Levels there is the RSPB's Newport Wetlands National Nature Reserve as well as the Gwent Wildlife Trust's wildflower meadows at Great Traston and Magor. In the context of the habitats required by the bees such nature reserves are invaluable but ultimately inadequate in their scale and isolation.

It is therefore the management of the whole landscape of the Levels that will determine the survival, or otherwise, of the Shrill Carder Bee. The countryside on which the bees depend on the Levels is made up of numerous landholdings with an abundance of drains interconnected by seawalls and interspersed within regenerating industrial sites. A light touch to pasture management helps provide forage for the bees throughout their long season. The survival of the Shrill Carder Bee and numerous other pollinators is dependent on countless individual management decisions such as when to cut a thistle patch or how to rest a horse paddock.

Current reinvestment in the old steelworks east of Newport and other industrial sites offers opportunities to sustain flower-rich habitats. The misnomer 'brownfield sites' as applied to these habitats detracts from recognising their importance for insects. With abundant bare ground and uncut flowery margins a brownfield may be far richer in nesting and foraging sites than the surrounding farmland. With skill and goodwill the redevelopment of redundant heavy industrial sites can help contribute to the natural green spaces of the future.

Binding the whole network together is the seawall. At a total length of some 35km the seawall grasslands of the Gwent Levels represent a substantial area of potentially bee-friendly habitats. There are stretches of clover-rich grassland and numerous foraging grounds with patches of thistles and Wild Teasel *Dipsacus fullonum*. Sympathetic grazing and mowing regimes combined with imaginative restoration when the seawall is re-engineered will deliver that potential. The seawall connects to the tidal outfalls and through them to the entire network of watercourses and fields. Such interconnections provide the routes along which young queens may search for new nest sites and so maintain the balance between local extinction and colonisation.

The treadmill

In his 2001 review of the history of sea defences the geo-archaeologist John Allen observed that the costs of safeguarding land taken from saltmarshes grows exponentially over time, a treadmill unlikely to have been anticipated when the process began.

There are consequences when a saltmarsh is excluded from the reach of the tide. As soon as a wall is completed the flow of sediments that built that land is abruptly cut off. At first the arrested development is inconsequential but over time it becomes increasingly problematic.

At Slimbridge the Wildfowl and Wetlands Trust nature reserve occupies land progressively embanked from the estuary since the medieval period. The most recent seawall was built in the early 19th century to take advantage of the growth of saltmarsh into the tidal river. Over successive years the estuary has continued to deposit silt and the unembanked marsh has continued to grow.

Slimbridge's tidal saltmarsh, the Dumbles, continues to change. A sandbar that recently accreted offshore is becoming stabilised through colonisation by Common Cord-grass *Spartina anglica* to form a new island. Should that trend of accretion and stabilisation continue

The Severn at Slimbridge.

then the Dumbles and the island will soon merge. Not only have the Dumbles grown outwards but they have also grown vertically. The consequences of this effect over time are illustrated in John Allen's study of 1986, which shows that land taken from the intertidal in the 19th century now lies 40cm below the surface of the adjacent saltmarsh. Moving further inland, that 19th-century land surface adjoins marsh embanked in the medieval period, the medieval land lying 70cm lower than its neighbour. Slimbridge's landscape demonstrates a general rule in embanking saltmarshes; the longer land has been embanked then the lower the level of that land is relative to the tide.

Any low-lying land behind a seawall is vulnerable to flooding. Freshwater flooding will happen whenever water enters that land faster than it can flow out to sea. These floodwaters gather at the lowest point and can only be discharged using gravity when the tide has ebbed sufficiently to create a gradient for the water to flow. If there is an insufficient gradient, or the discharge is curtailed by the rising tide, then that water is trapped. These circumstances naturally occur in unmodified estuary systems where tides impede the flow of freshwater. Such natural impoundments help explain the origin of peatlands that have developed on the periphery of tidal rivers and the

The Somerset Levels from Brean Down.

landward reaches of saltmarshes. On the Severn these are particularly well developed in the Back Fens of the Gwent Levels and the Peat Moors of the Somerset Levels. With sea-level rise and climate change these low-lying moors are increasingly vulnerable to floods.

Disastrous floods in the historical past were assigned to the judgement of God; current events trigger an equally irrational and predictably unedifying spectacle. Senior politicians and journalists are swept in on the flood to demonstrate their concern by casting round to attribute blame. Any attempts to offer informed analysis are scorned and displaced by heartbreaking stories of suffering. Simplistic solutions are promoted which meet immediate expectations yet, in more considered times, are known to be ineffective. Political promises may comfort the distressed but, no matter what short-term measures are applied, if the land is below sea level then it will flood again; ultimately the force of the tide is irresistible.

Responses to risk

With ingenuity and an unrestricted budget it is possible continually to enhance the scale of engineering to keep pace with tidal changes. Seawalls are heightened and broadened, cut-off channels and elevated drains are built to divert freshwater with ever-elaborate sluices to discharge into the estuary. Where gravity is insufficient then pumps will force water to flow uphill and drains can be dredged in the hope of hurrying that water along. History teaches us that the risks and associated costs of maintaining seawalls rise exponentially over time and that, sooner or later, seawalls fail.

The seawalls embanking the Severn are currently inadequate to cope with predicted environmental change. The Environment Agency (2011) has applied conservative models of sea-level rise and climate change which indicate that over the next century local tides will gradually rise to be a metre higher than today, with peak river flows into the estuary increasing by up to 20 per cent.

The 2011 review identifies 197,500 residential and commercial properties currently benefiting from existing defences. The cities of Cardiff and Bristol have spread from their historical centres over land that is at or below the level of the tide. A study of the Gwent Levels by Stephen Rippon in the 1990s found that 39 per cent of the former saltmarshes had been built over, with another 12 per cent allocated for further growth (Rippon 1997). At the time of writing there are plans

A consequence of living below sea level; tide gates on the River Axe, Somerset.

to redirect the M4 motorway across the Levels which, if built, would increase that proportion and inevitably attract further urban growth. Strategic infrastructure including motorways, railways, power stations and sewage works have all been constructed on land below the high tide. In 2011 the risk to property and infrastructure of seawall failure along the Severn was costed at £5 billion.

Historically most seawalls were built to improve the agricultural value of the land, and farming remains the most extensive industrial use of the Levels. Flood-risk modelling of the Severn indicates that a complete failure of the seawalls would lead not only to the flooding of the 35,000ha of former intertidal but also a potential additional 65,000ha of adjoining low-lying land.

Rising sea levels can put wildlife at risk, particularly where previous engineering interferes with the landward migration of the intertidal. The impact of seawalls on saltmarshes is not wholly predictable. Current models suggest that over the next 100 years the presence of current, and upgraded, coast defences along the Severn will result in the loss of between 1,500 and 3,500ha of intertidal habitat; fortunately the realignment of upgraded seawalls offers opportunities to rejuvenate the natural processes of the intertidal.

The response of wildlife to recent changes in flood defences can be seen at Aylburton Warth, near Lydney on the shore of the Forest of Dean. Here a realignment of the sea defences in 2013 has returned

39ha of conventionally farmed land to the intertidal. Within the first two years the revitalised marsh took on much of the character of the neighbouring warths. A fine-grass sward developed with an abundance of English Scurvygrass *Cochlearia anglica*, which despite its name is related to cabbage. Maintaining opportunities for colonisation by saltmarsh species has been secured through livestock whose grazing and poaching prolongs the availability of bare ground for tide-borne seed. The early establishment of this relatively short turf is proving attractive to wintering Eurasian Wigeon *Anas penelope*, Golden Plover *Pluvialis apricaria* and Lapwing *Vanellus vanellus*. The extension of the grassland also increases the potential breeding territories of the waders already established on the adjacent warth. Having rebuilt the seawall on its new alignment, the old wall has been left to be taken by the tide. Each tide scours the face of the marsh but the new alignment of defences is more readily maintained than the old. The retained sections of the old seawall provide roosting sites above the reach of the high tides. It is here that hundreds of wintering Curlew *Numenius arquata* gather in safety until the ebb exposes their feeding grounds.

Elsewhere in the estuary changes are occurring that have the potential to generate successors to threatened grazing marsh habitats. Upstream of Aylburton the low seawalls at Saul Warth, by Frampton, are vulnerable to overtopping at high tides. The land protected by the wall is currently grazed but until recently it was farmed for potatoes amongst other crops. Today livestock are free to wander along the foreshore, across the seawall and over the former cultivations. Naturally rising ground forms the landward limits out of which flows a series of vigorous freshwater seepages. This combination of circumstances has supported the emergence of a large seasonal pool of mildly brackish water, an analogy to the environment of grazing marsh ditches but generated by natural topography rather than engineering. The brackish wetlands of Saul Warth are as young as the new marsh at Aylburton. Time and chance will determine what insects discover the Warth and find it to their liking.

The United Kingdom is party to international treaties and has passed national legislation to safeguard internationally important habitats like the Severn estuary from loss. The scale of predicted habitat loss requires changes of a far greater scale than those at Aylburton and Saul Warth to meet those obligations. In the 1950s, seawalls were built along the western shore of the tidal Parrett where it

Restoration of the tidal
Parrett at Steart.

flows into the Severn at Bridgwater Bay. What had been a flood-prone
reach of the estuary on the Steart peninsula became, for a few decades,
suitable for dairying and arable farming. The cost of maintaining the
seawalls became progressively difficult to justify and by 2002 the local
community were exploring options to safeguard their homes whilst
finding a fresh approach to coast defence.

In 2014 the new flood defences at Steart became operational. The
Environment Agency realigned the seawalls to focus on protecting
people, their homes and businesses. The old seawall has been
breached to create an intertidal basin of some 200ha set within a
broader landscape of brackish marshes and lagoons. The majority
of the new wetland, under the management of the Wildfowl and
Wetlands Trust, will remain farmed with plans for cattle and sheep
to graze across the site.

The value to wildlife in grazing the marshes is illustrated by recent
changes in the estuary of the neighbouring River Axe. Here the Avon
Wildlife Trust's Walborough Nature Reserve includes embanked
marshes together with marshes fully open to the tides. Cattle
graze the wetlands, moving with the season through the adjoining
limestone grassland, with its Honewort *Trinia glauca* and Somerset
Hair-grass *Koeleria vallesiana*. The Walborough marshes are similar

to those upstream, being grassy and supporting Bulbous Foxtail and Strawberry Clover. They are however far richer than typical Severn saltmarshes, being diversified by many more demanding species including Sea Wormwood *Artemisia maritima*, Parsley Water-dropwort *Oenanthe lachenalii* and Sea Clover *Trifolium squamosum*. In contrast, immediately upstream from Walborough are the Bleadon Levels, recently restored to the intertidal. Their alignment means they are inaccessible to livestock; the net result is a rather lank mat of coarse vegetation devoid of the structural and biological diversity of its grazed neighbour.

Under the Wildfowl and Wetlands Trust's management the newly rejuvenated Steart Marshes have the best chance to develop without a few species becoming dominant to the exclusion of others. The subtle approach to livestock management perfected at Slimbridge provides a model to guide the development of Steart now that the walls have been breached.

The tendency to think of seawalls as permanent structures has been described by Allen & Rippon (1995) as an assumption untenable not just in the Severn Estuary. Sea defences are interruptions to powerful natural processes; they suspend the landward migration of the intertidal but only for as long as the wall holds.

Capital marsh country

L ondon grew by, and now grows over, the saltmarshes of the Thames estuary. The marsh country has provisioned the city with the basics of life, with food and medicines, with inspiration for artistic endeavour and as a gateway to the world. Tidal waters flow through the heart of the capital.

The peregrinations of Thomas Johnson

In the library of Magdalen College, Oxford, there are two volumes, barely more than pamphlets, describing the Thames estuary during the early years of the 17th century. In *Iter Plantarum* (1629) and *Descripto Itineris* (1632) Thomas Johnson records two journeys by a group of apothecaries out of London into the countryside of north Kent. Behind the respectable cover of Latin texts are joyful recollections of companionable botanising. Fortunately in 1972 a translation was published by John Gilmour so we may all share in the friends lolling in brewers' drays, dining lavishly and on occasion enjoying suppers 'seasoned not so much with variety of dishes as with harmless jests'.

A suspicious official on the Isle of Sheppey was advised by the party 'We are devoted to the study of science and material resources of medicine. That is why we have come to this place to discover the rare plants that grow in your island'. This rational response, combined with unctuous flattery, left them on the most convivial of terms.

Johnson chose to record his discoveries not as a single account but through grouping species by when and where they were found. Their trip to the shore by Queensborough Castle resulted in two lists, one of saltmarsh and the other 'on the shore itself'.

OPPOSITE PAGE:
Marsh country on
Sheppey, Kent.

Yellow Horned-poppy on a marsh's shingle shore.

Glassworts *Salicornia* spp. and Annual Sea-blite *Suaeda maritima* are annual plants of open intertidal sediments and were recorded by Johnson at Queensborough. They were accompanied by a suite of perennial species from higher up the marsh including Small Cord-grass *Spartina maritima*, Sea-purslane *Atriplex portulacoides* and Golden-samphire *Inula crithmoides*. A separate list for Queensborough's shore reflects the perennial vegetation of a stony beach with Sea Spurge *Euphorbia paralias*, Yellow Horned-poppy *Glaucium flavum* and Sea Kale *Crambe maritima*. This grouping of records can be read as an instinctive recognition of the constancy of certain species with their environment and with each other. It would be centuries until other botanists articulated the concepts of habitat and community; Johnson's accounts suggest that he had already started on that path of reasoning.

Mental pictures are readily summoned up by Johnson's evocative botanical snippets. The strandline by the Royal Docks is strewn with debris through which straggles a line of Sea Wormwood *Artemisia maritima*. Grassy salterns at Cliffe offer a gentler scene softened further through the pinks of Strawberry Clover *Trifolium fragiferum* and Marsh-mallow *Althaea officinalis*. The discrete bounds of Westgate Bay encompass a complex of dunes and saltmarsh grading to the brackish grounds of Sea Club-rush *Bolboschoenus maritimus* and

Wild Celery *Apium graveolens*. Naturalists still share their impressions of favoured places through citing a few choice species and a little topography.

Most of Johnson's journeys were undertaken on foot, with boats carrying them from London and then between the marsh islands of Sheppey and Grain. Not every excursion proved valuable:

> *and after leaving the little ship ... we walked five or six miles without seeing a single thing that could give us any pleasure ... In the heat of the day we were tormented like Tantalus with a misery of thirst in the midst of waters – they were brackish. We were equally afflicted with hunger in that inhuman wilderness where there was no town within reach, no smoke to be seen, no barking of dogs to be heard, none of the usual sights of habitation by which we could arouse our fainting spirits to any breadth of hope.*

There was, however, hope of pleasure in a diversion down to Faversham. The marshes there were known as the source of a great rarity much sought after by herbalists. This was Hog's Fennel *Peucedanum officinale*, a long-lived spicily sulphurous member of the

Elmley Marshes, a family-run National Nature Reserve, Kent.

carrot family. When Johnson and his friends visited in August the herb would have been in full flower with massed umbels of inconspicuous blooms held up to eye level. This is not a species one has to scrabble for. As summer progresses the herbalists' prize becomes progressively easy to find. The bulk of the plant is a neat knee-high filigree mound which on maturity takes on the tones of old gold. An experienced eye will track along the shoreline, just at the fullness of the flood, to find where grasses change from glaucous Sea Couch *Elytrigia atherica* to the conventional field greens of fescues and bents *Festuca* and *Agrostis* spp. It is on this transition where Hog's Fennel forms its metallic band and towers above the other components of the marsh.

The pungent aromatics that help protect Hog's Fennel from grazing animals are concentrated in the deep tap-root. Apothecaries extracted the gummy juices by either digging out the whole root, with much labour and sacrifice, or tapping the sap rising from the living plant. Later in the century Faversham would be recognised as a source of Hog's Fennel by the herbalist Thomas Culpeper (1653) who recommended it for a host of complaints ranging from lethargy, frenzy and gangrene to shortness of breath and wind.

Johnson's trips were purportedly about studying science and the material resources of medicine. This may well have been so, but he

Mature stands of Hog's Fennel by the Medway, Kent.

did not feel it necessary to capture those particular matters amongst his descriptions of botany and bonhomie. Reading his accounts, one cannot help feel that, in addition to the worthy reasons, Johnson and his friends revelled in the naturalists' pleasure in pure curiosity.

It had been Thomas Johnson's intention to prepare a book describing all the wild plants of Britain. The project proved impossible, with the breakdown of relations between the King and Parliament leading to civil war. Johnson died of wounds at the siege of Basing House.

Fellow apothecaries

Thomas Johnson's professional life and that of his travelling companions was governed by the Worshipful Society of Apothecaries. The Society of Apothecaries grew out of the guilds that from the 12th century had controlled London's trade in herbs, spices and groceries. Over time the apothecaries grew to regard themselves as superior to mere grocers as they manufactured, prescribed and dispensed medicines. Having convinced the King of their case for independence, and to the lasting displeasure of the grocers, the company was established by Royal Charter in 1617.

The raw materials of the apothecaries' medicines were plants, many of which grew around the city. It was important for apothecaries to know how to find plants and how to tell them apart. Misidentification may prove an expensive, even fatal, error. An essential element of the training of an apothecary was therefore botany. As Johnson advised in his preamble to *Descripto Itineris*:

> For the doctor relies on the druggist and the druggist on a greedy and dirty old woman with the audacity and the capacity to impose anything on him! So it often happens that the patient's safety depends on the herbal knowledge of an ignorant and crafty woman.

The Society of Apothecaries survived the civil war and resumed their work. The records that followed the restoration of the monarchy are invaluable, but Johnson appears to have been singular in his recording of groups of plants rather than individual species. Even so, single species accounts can still provide useful insights for the modern naturalist, particularly where the species is exacting in its ecological requirements.

Dougie Kent's *Historical Flora of Middlesex* (1975) includes records of saltmarsh plants from early 18th-century London, notably

English Scurvygrass *Cochlearia anglica*, Wild Celery, Sea Club-rush and Marsh Sow-thistle *Sonchus palustris*, respectively from the Isle of Dogs, Westminster, Battersea and Chelsea. Each record indicates something of the habitats present at that time and how the land was being managed. Saltmarsh habitats, particularly those at the brackish end of the spectrum, were common and widespread. Some of the records, such as those of Joseph Andrews, contain sufficient clues to tentatively re-create their historical landscape setting. Andrews was apprenticed to John Field, apothecary of the Bell in Newgate Street. Together they gathered a specimen of Sea Clover *Trifolium squamosum* 'by the Thames side near the Earl of Peterborough's Palace'. The reference to the palace assists in placing the record in the vicinity of Parson's Green on the Fulham Road.

Walking down the Fulham Road today, it is difficult to envisage the landscape enjoyed by Andrews and Field. Sea Clover is associated with brackish grassy marshes cropped by cattle or cut for hay; it is a plant of big landscapes and open skies. Today the notion of periodic inundation by saltwater in such a fashionable part of west London is inconceivable, improbable but not impossible.

Further insights into London's saltmarshes are offered by records of Triangular Club-rush *Schoenoplectus triqueter*. This specialist of swamps by tidal rivers was first recorded by William How in 1650

A short section of Triangular Club-rush.

by the horse ferry in Westminster, then again by James Pettiver in 1695 between How's site and Peterborough House. The club-rush attracted may other botanists and so there is a sequence of records through the following centuries. Being labelled a rush implies a rather dull thing, a moniker that does disservice. Triangular Club-rush is a strikingly angular plant appearing to be manufactured from a shiny green sugar-cube, carved into a three-sided pyramid and then stretched until it reaches above eye level. The three equal sides give the implausibly slender structure rigidity, a vegetal shard with a perfect apex. Flowers emerge pressed to the stem like a corsage of stubby brown sausages precisely elevated to bloom above the tide; here is naturally engineered elegance.

Triangular Club-rush is now lost to Britain as a species of the wild. Plants from cultivated stock persist on the Tamar in Cornwall; the former natural colonies eventually yielded to the spread of tidal reedbeds following the cessation of riverside grazing. Triangular Club-rush may be gone but its hybrids persist elsewhere in southern England and one can travel to the tidal Shannon to see the unalloyed species in a natural setting. This depressing loss relates to the very address of the government ministry charged with caring for the environment, Defra at Horseferry Road. The irony is compounded by the last Thames population succumbing in 1946 to a major reconstruction of the river-wall which destroyed the last patch of estuarine mud on which it grew. That ill-fated mudbank was within sight of Kew Gardens, the international centre for plant conservation.

Maligned marshland

The Thames estuary has become somewhat of a visual cliché in TV dramas. London's marshes are a convenient location to imply grime and feral thuggery. All too often the estuarine backdrop provides the setting for a gritty crime drama or the weary dénouement of a tale of spies.

The origin of this stereotype can be sought in the novels of the late 19th century. Charles Dickens knew intertidal mud since his boyhood on the marsh-bound city of Portsmouth, and is well known for his much quoted 'I was born at Portsmouth, an English seaport town, principally remarkable for mud, Jews and Sailors'. His greatest novels, however, are synonymous with the less salubrious side of the capital.

Great Expectations (1861) opens with young Pip crossing the Thames marshes to be waylaid by Abel Magwitch in a graveyard. Dickens describes the marshes in all their moods and seasons as being a happy home to Pip. They are his familiar landscape; he recalls 'Ours was the marsh country, down by the river'. The events and personalities Pip encounters are terrifying, but the emotions arise from the people, not the place.

In contrast, the estuary dominates the opening shot of David Lean's 1946 film adaptation of the novel. The child Pip is dwarfed by the immensity of a marsh at high water under the half-light of a louring sky. Pip hurries on, he passes a bend in the seawall, hesitates at an empty gallows and runs out of shot. Our view is from just above head height, we follow him in an unblinking single take, we watch unnoticed.

The landscape exudes menace, the camera conjuring harsh angles and bold contrasts from the Medway flatlands.

In Dickens' last complete work, *Our Mutual Friend* (1864), two colleagues consider the quality of London on a cold spring day: 'such a gritty city; such a hopeless city, with no rent in the leaden canopy of its sky; such a beleaguered city, invested by the great Marsh Forces of Essex and Kent.' Language has evolved to deprive us of the full meaning of the conversation beyond that emphasised in capitals. The investiture entertained by Messrs Lightwood and Wrayburn was the pervasive force that saturated the city with its sawing east winds. In all its grandeur the city cannot resist the great Marsh Forces.

The juxtaposition of marsh and city was more distinct in the closing decades of the 19th century than it is today. The marsh skylines had yet to be punctuated by towers and lights, nor were the flatlands built over. London was then the capital not only of Britain but of an empire, the river and its marshes their gateway. Joseph Conrad's narrator of a journey into the *Heart of Darkness* (1899) recounts his tale from a tide-bound yawl on a sea-reach of the Thames. The preamble to Marlow's recollections of his imperial adventures is a speculation on the anxieties of ancient Romans entering the Thames for the first time. Such a journey would take a civilised man into an incomprehensible, detestable dark place. We know of the reluctance of the Romans in advancing into Britain from the histories of Cassius Dio as translated by Cary (1914). Platius, a commander at the time of Claudius, 'had difficulty in inducing his army to advance beyond Gaul. For soldiers were indignant at the thought of carrying on a campaign outside the limits of the known world'. Having persuaded his troops to land,

> the Britains retired to the river Thames at a point where it empties into the ocean and at flood-tide forms a lake. This they easily crossed because they knew where the firm ground and the easy passages of the region were to be found; but the Romans in attempting to follow them were not so successful.

To imperial minds, classical and Victorian, the wild places would have to be subdued before the noble cause of progress could flourish.

As it reaches the sea the Thames is joined by lesser rivers to form a complex of estuaries from Thanet to Felixstowe Ferry. Progressing through the Essex creeks of the Crouch, Colne and Blackwater, one eventually reaches Mersea, the island home of the Reverend Sabine Baring-Gould. Whilst now fallen from fashion, Baring-Gould was eminent in his day. He was a man of prodigious energy: he wrote the popular hymns 'Now the day is over' and 'Onward Christian soldiers'; he collected folk songs; he published over 1,200 works including the lives of the saints and a study of werewolves; and at the age of 34 he married a girl half his age who went on to bear him 15 children.

Baring-Gould titled his novel of 1880 *Mehalah, a story of the salt marshes*. The marshes around Mersea were appropriated and the name of a local girl adapted for this tale of appalling sadism. Victorian readers were thrilled by the relentless descent of the independent-minded heroine through robbery, dispossession and forced marriage to be murdered whilst unconscious, chained to her monstrous husband as he scuppers their boat. For those less enamoured of melodrama the novel has lyrical interludes celebrating the natural beauty and rural economy of a saltmarsh parish. The overture to the dreadful tale sets Mehalah's home on Ray Island, a marsh farm by the causeway to Mersea.

A more desolate region can scarce be conceived, and yet it is not without beauty. In summer, the thrift mantles the marches with shot satin, passing through all gradations of tint from maiden's blush to lily white. Thereafter a purple glow steals over the waste, as the sea lavender bursts into flower, and simultaneously every creek and pool is royally fringed with sea aster. A little later the glasswort, that shot up green and transparent as emerald glass in the early spring, turns to every tinge of carmine.

When all vegetation ceases to live, and goes to sleep, the marshes are alive and wakeful with countless wild fowl. At all times they are haunted with sea mews and roysten crows; in winter they teem with wild duck and grey geese. The stately heron loves to wade in the pools, occasionally the whooper swan sounds his loud trumpet, and flashes a white reflection in the still blue waters of the fleets. The plaintive pipe of the curlew is familiar to those who frequent these marshes, and the barking of the brent geese as they return from their northern breeding places is heard in November.

As with Pip's encounters with Magwitch, and with Marlow's journey up river, it is in the people, not the place, where the horror lies.

A mixed diet

Today the saltmarshes of the Thames are the shattered remnants of a great estuary. The landward reaches of the marshes have been comprehensively modified through enclosure behind seawalls. On the Essex shore alone Gascoyne & Medlycott (2014) account for over 22,000ha of saltmarshes being embanked, of which only 6,500ha survive as grazing marshes. From the 1930s onwards Thornton & Kite (1990) have traced the history of livestock being replaced by ploughland, and then ploughland by homes and industry. The capital grows seaward on its veneer of engineering; the splendour of Canary Wharf stands on tidal ooze.

Thames estuary seawalls have been dated from the 12th century; the original structures were modest in scale and there are numerous records of them being overtopped. The driving force behind these medieval embankments appears to be dairying, for sheep rather than cattle. Milk from saltmarsh flocks was converted into hard truckles of cheese in numerous Wicks. Wick, the local name for dairies and cheese makers' sheds, is still commonplace around the Essex coast. There was a ready market for the cheesemongers just a tide or two upstream; in the victualling yards marsh cheeses were known for their keeping qualities, well suited to a long sea voyage. Further upstream the city's waged

A view across Fingringhoe Wick from Essex Wildlife Trust's nature reserve.

poor found them good value, if something of an acquired taste. It was not until the late 16th century that growing affluence shifted enterprises on the marshes to the higher-value protein of beef.

An historical landscape of seawalls can be walked to the east of Fobbing on the Essex shore. In a marsh enmeshed with pylons Least Lettuce *Lactuca saligna* grows on an ancient bank of clay. Least Lettuce is closely related to the salad crop but unlike that vegetable it is quite inedible. A milky acrid latex flows through every vein of the crisp, juicy leaves. Even in closely grazed pastures the cattle avoid their distasteful rosettes. Most specimens are small, hence 'Least', but in exceptional circumstances Fobbing's seawalls can sport willowy frameworks of waist-high stems. In high summer each plant carries hundreds of yellow flowers, dandelion-like in character, but never as showy and always closed-up by noon.

In Britain Least Lettuce is an exceptional rarity, being confined to the Thames estuary and Rye Harbour (Sussex), but it is widely distributed across continental Europe from the Mediterranean to central Russia.

TOP: Fobbing Marsh.

ABOVE: Least Lettuce.

The east of England shares elements of that continental climate with the lowest rainfall in Britain and some of the highest hours of sunshine. Impending autumn brings dry breezes that fluff up the dandelion-clock seeds and assist their dispersal across the landscape. Such a strategy is effective where the seeds float freely across open marshland; such powers are inconsequential if there is nowhere left to disperse to.

There was a time Least Lettuce was widely distributed around the coast of south-east England. It is one of a suite of species of the upper limits of saltmarshes and tidal rivers within the zone of occasional flooding. At Fobbing the seawall on which it persists has been superseded by engineering aligned closer to the river. Redundancy as a primary coast defence means the wall has not been upgraded and a modicum of broken ground is acceptable. Cattle wander over Fobbing's landlocked embankments where their slithering hooves create spaces for germination. Elsewhere around the estuary the loss of grazing marshes, upgrading of seawalls and the exclusion of livestock have turned what was always a local species into one of Britain's rarest plants.

The botanical treats of the Thames include some of Britain's least glamorous species; large muddy estuaries are places to enjoy fleshy chenopods. Chenopod translates from the Greek as 'goosefoot' in recognition of their association with places where geese were raised. The family includes species caricatured as dunghill superweeds as they do extraordinarily well in the crusts and liqueur of muckheaps. In the wild, such conditions are created where large mammals congregate near water; their dung raises fertility and their trampling maintains bare ground. In modern landscapes such circumstances are rare, increasingly so as livestock management seeks to be compliant with inflexible environmental regulations.

Opposite Fobbing, on the Kentish shore, traditional grazing marshes persist around the Swale and Sheppey. Whilst divorced from twice-daily tides, some former creeks retain their saltiness, the brine concentrating with the droughts of summer. Saltmarsh Goosefoot *Chenopodium chenopodioides* is the doyen of these muddy fleets.

Saline wetlands experiencing extreme seasonal fluctuations are ideal for annual plants. Saltmarsh Goosefoot is accompanied in such a habitat by its close relatives, Fat-hen *Chenopodium album* and Red Goosefoot *Chenopodium rubrum*. As soon as the fertile mud is exposed to air there is a rash of seedlings, followed by successive generations

Saltmarsh Goosefoot.

as the water subsides. Rings of mealy leaves trace the season's falling waters, their concentric lines punctuated by the hooves of wandering stock. Fertile mud and full sunlight are ideal conditions for rapid growth; in a few weeks flowers emerge and seed is set. By September the mud shines red as plants senesce, a brief blaze of colour before the onset of rain and winter floods.

As long as archaeologists have recorded human diets chenopods have been on the menu. The extended family includes a variety of species including the oraches *Atriplex* spp., beets *Beta* spp. and glassworts *Salicornia* spp. Domestication has brought many chenopods into the kitchen. Beet *Beta vulgaris* ssp *maritima* is the native species improved on by plant breeders to create 'Swiss' chard (subspecies *cicla* var. *flavescens*), together with beetroot and white sugar refined from sugar beet (subspecies *vulgaris*). In recent years fashion has favoured chenopods with native 'samphire' *Salicornia* spp. decorating seafood dishes and the Andean *Chenopodium quinoa* promoted as a super-food.

Saltmarsh chenopods have been described to me by foragers as being 'a bit like spinach', which can be loosely translated as being green and unlikely to poison you. Today the most commonly grown commercial forms of spinach are the Asian *Spinacia oleracea* and *Atriplex hortensis*. These delicate members of the family have replaced the rather coarse Good-King-Henry *Chenopodium bonus-henricus*, a once commonplace pot herb. Geoffrey Grigson's *Englishman's Flora* (1955) describes the young shoots of this vegetable as neither very pleasant nor unpleasant, which is not much of a gastronomic recommendation. Good-King-Henry was introduced into Britain as a foodplant but succeeded in jumping the garden fence into the wild. It persists in waste places, where it is treated as a curiosity, an alien but not invasive. In contrast, Fat Hen, another spinach substitute, is native to our saltmarshes but offers equally questionable culinary qualities. Local names from around England testify to its preferred habitat: Dirtweed (Lincolnshire), Dungweed (Gloucestershire) and Dirty Dick (Cheshire). Today our ready access to a varied diet means we need no longer opt for such uninspiring choices.

Comings and goings

Rarest of all of the Thames's chenopods is the Pedunculate Sea-purslane *Atriplex pedunculata*, another annual of the upper marsh. Through the early 20th century Pedunculate Sea-purslane declined throughout its British range. Specimens were collected in 1932 from Freiston on the Lincolnshire Wash, a shore then being embanked for farming. Ted Lousley, professional banker and renowned amateur naturalist, saw the species in 1938 in Suffolk at Walberswick. Circumstances did not encourage further survey as shortly after his visit the East Anglian coast was fortified with wire and mines to deter invasion. Lousley was the last British botanist to see the species for nearly half a century.

As one of its co-discoverers observed, the rediscovery of this presumed extinct species was as much a matter of luck as judgement. In 1987 Simon Leach and his colleagues from the Nature Conservancy Council were surveying the marshes of the Essex coast. By September they had reached Foulness, a landscape of farmed creekland and military ranges conjured by seawalls from an archipelago of mud. By this late season the chenopods were mature, the most striking of which bore the seedcases of *pedunculata* whose pendulous form renders it unforgettable (Leach 1988).

Whilst historically distributed around the coast of south-east England, Pedunculate Sea-purslane had never been previously recorded from the Essex reaches of the Thames. It was possible the species was a long-standing member of the local flora unrecorded due to access restrictions and obscurity. Some marshes along this coast were cattle grazed in the late 1980s but this did not extend to the purslane colony, which

Flowers and young fruit of Pedunculate Sea-purslane.

was surrounded by rank grass. The population was vulnerable to being overgrown and by the mid-1990s was in serious decline. Its precarious foothold in the wild justified bringing material into cultivation where it was found to grow freely from seed. Wyatt (2002) reported that attempts were made to establish new colonies elsewhere on the Essex coast but without success. The failures were attributed to having selected the wrong habitat, with no consideration to whether it was the right habitat but under the wrong management. Pedunculate Sea-purslane will remain a challenge to conserve as long as it is a species of another time and place, of free-ranging cattle and trodden creeks. Without such ecological processes being restored to the landscape of the Thames the purslane will remain dependent for its survival on repeated interventions by conservationists. Seeds help plants survive such periods of adversity; seedbanks are investments against an uncertain future.

Unlike plants that can set seed, some invertebrates deal with changing environmental conditions through their mobility. Coastal grazing marshes that support waterbodies of various salinities can be exceptionally rich in dragonflies and damselflies. In being close to the continent the established species of the Thames marshes are regularly supplemented with migrants, which add to their diversity. In 1946 the Dainty Damselfly *Coenagrion scitulum* was discovered breeding in a grazing-marsh pond near Hadleigh in Essex. Dainty Damselflies have a world distribution, taking in much of Africa, southern Europe and central Asia. They are not particularly specialised in their habitat requirements, having a preference for sunny standing-water with submerged plants such as hornworts *Ceratophyllum* spp. and water-milfoils *Myriophyllum* spp. Their catholic tastes extend to salinity; most populations occur in freshwater but they will successfully breed in weakly brackish conditions. Whether the damselfly had colonised the Essex coast unrecorded during the war years was unknown; the single locality suggested a recent arrival.

On the night of 31 January 1953 extremes of wind and tide drove a storm surge into the narrows of the North Sea. This exceptional tide overwhelmed sea defences across eastern England and the Netherlands, resulting in saltwater reasserting its presence over 360,000ha of former marsh. Flooding and fatalities were greatest in the Netherlands, where some 1,800 people died. Close by the Dainty Damselfly site, the sea broke through into Canvey Island where 58 people were drowned. The inundation and loss of the Dainty Damselfly pond is a footnote to this tragedy.

Dainty Damselfly.

For seven years the Dainty Damselfly was considered a British species before being added to the list of extinctions. 'Extinct' is an emotive word and needs to be used with caution. Dainty Damselflies ceased to breed in Britain in 1953 but on a world scale they remained widespread. Whilst not migratory, the damselfly is opportunistic in colonising new sites when circumstances are in its favour. Boudot (2014) has tracked its movements since the mid-1990s when the southern European population started heading north. By 1995 the damselfly had reached north-east France and into Luxemburg in the following year. Belgium was colonised in 1998, Switzerland in 2001 then northwards to the Netherlands in 2003. The open sea proved no barrier to

establishment in the Channel Islands in 2009 and onwards to the Thames marshes in 2010. With the benefit of hindsight the 1946–53 population was not so much an extinction as a prelude.

Survivors of the Canvey Island flood, 1953.

The current colonisation of Britain by the Dainty Damselfly may persist; the international trend is in its favour, as is the experience of other dragonfly species. Scarce Emerald *Lestes dryas* is a resident of the Thames marshes and as recently as 1983 was feared by Cyril Hammond to be extinct due to habitat loss and pollution. This nominally 'Scarce' damselfly is now doing so well in the Kent marshes that it may require a new English name. Similarly, recent years have transformed the Southern Migrant Hawker *Aeshna affinis* from a prized itinerant to a breeding species of the Essex marshes; the migrant has become a resident and our communities are enriched. What dragonflies lack in resilience they make up for in their ability to disperse; in the long run everything must move to survive.

The legacy of the Solent's saltworks

Until the rise of coal as a cheap source of power, the Western Solent in Hampshire was the home of many heavy industries. An abundance of timber and charcoal provisioned the iron foundries and shipyards of Nelson's Navy. Long hours of sunshine and low rainfall gave a natural competitive advantage to manufacturing salts on an industrial scale. Through the course of the 19th century these heavy industries became redundant as other regions of Britain developed more efficient enterprises. Most of this post-industrial landscape has been recovered by nature and is now renowned for its beauty as part of the New Forest National Park.

In August 1833 the young Charles Darwin took a trip into the mountains of Patagonia. Riding out from the town of El Carmen, he visited salt lakes where Chilean Flamingos *Phoenicopterus chilensis* lived in abundance around a crystalline shore. Darwin was on the outbound leg of a journey that would take nearly five years and which in time stimulated his theory of evolution through natural selection. His journal of that journey was published in 1839 and is better known today as *The Voyage of the Beagle*.

In a footnote to his account of that day Darwin recalled a description he had read of a crustacean, a shrimp, occurring in countless numbers in the saltworks of Lymington on the Hampshire coast. The shrimp was found there in brine pans in which the fluid had attained considerable strength through evaporation; a concentration of a quarter of a pound of salt to a pint of water.

OPPOSITE PAGE:
Keyhaven marshes.

This remarkable creature prompted him to contemplate on the nature of life:

Well may we affirm, that every part of the world is habitable! Whether lakes of brine, or those subterranean ones hidden beneath volcanic mountains – warm mineral springs; the wide expanse of the ocean; the upper reaches of the atmosphere; and even the surface of perpetual snow; – all support organic beings.

Darwin's recollection was of the Brine Shrimp *Artemia salina*. He was certainly not the first traveller to marvel at its ability to thrive in the most extreme of hostile habitats. The earliest record of these shrimps has been extracted by Asem (2008) from an anonymously authored geography book that described them from Lake Urmia in Iran. Written in Persian, *The Limits of the World from the East to the West* has been dated using the Western calendar as from 982 AD.

The honour of first recording Brine Shrimps in Britain goes to a Dr Schlösser who in 1756 provided a description and drawings from saltworks on the Hampshire coast of an 'insect peu connu'. Without doubt the workers would have known the species long before the Doctor.

An adult male Brine Shrimp.

Brine Shrimps are confined to waterbodies with salinities significantly stronger than seawater. Such habitats occur naturally,

even in Britain where rainfall exerts a diluting influence. Roger Bamber (2010) has reported that the concentration of salts in isolated bodies of seawater in eastern England can be enhanced by natural evaporation from 35 to 80 parts per thousand. In the 19th century the artificially concentrated brine of the Lymington saltworks was in the order of 100 parts per thousand. To reliably crystallise salt in a British climate it is necessary to boil off excess water. At Lymington skilled workers could control the process to precipitate different elements, so separating table salt (sodium chloride) from purgative Epsom salts (magnesium sulphate) and laxative Glauber's salts (sodium sulphate). Brine Shrimps played their part in the manufacturing process by filtering out impurities, particularly bacteria that could discolour the crystallising salt. Known to local workmen as brine- or clearer-worms, they were valued to such a degree that they were moved between tanks to accelerate the cleaning process, so creating an unlikely collaboration between wildlife and the chemical industry.

Lymington's saltworks grew to occupy most of the embanked saltmarsh between the town and the village of Keyhaven, 4km to the west. At its peak in the late 18th century the works supported nearly 200 boiling pans producing 6,000 tonnes of salt a year, estimated to represent about 10 per cent of Britain's salt production. By the early 19th century the industry was in decline and in 1866 it finally succumbed to the more efficient rock salt trade from the English

The last remains of Lymington's salt industry.

Midlands. Livestock proved more profitable than salt and so the works were cleared away and reverted to marsh.

From an unpromising origin the post-industrial landscape of Lymington's saltworks is remarkably rich in wildlife. There is nothing to see of the former wind pumps, boiling houses and evaporation ponds; the only standing monuments to Victorian enterprise are two barns. Open water is still abundant across the site, with a variety of lagoons under varying salinities. A pond previously used for purifying oysters has been colonised by the Foxtail Stonewort *Lamprothamnium papulosum*, here in one of its English strongholds. Other lagoons across the site support the Starlet Sea-anemone *Nematostella vectensis*, the scientific specific name commemorating it being first described from the nearby Isle of Wight, the Romans' Insula Vecta. The stonewort differs from the sea anemone in preferring brackish conditions, diluted to less than full-strength seawater. In contrast, the sea anemone will tolerate higher salinities, up to 40 parts per thousand, a useful adaptation should lagoon waters become concentrated over a droughty summer. Another denizen of Lymington's lagoons is a shrimp *Gammarus insensibilis*, unimaginatively dubbed the Lagoon Sand Shrimp. This is a predominantly Mediterranean species of the open sea that lives in brackish lagoons at the climatic limits of its range; like the sea anemone it is able to tolerate a limited increase in salinity over seawater.

Saltmarshes and brackish grasslands have recolonised the former industrial sites. The seawall that separates these marshes from the intertidal truncates the full sequence of habitats to the low tideline whilst enhancing the landward transitions. In common with many other saltmarshes in central southern England, the richest habitats are behind the seawalls. Cattle maintain muddy conditions where early succession species such as

ABOVE: Saltmarsh horseflies *Atylotus latistriatus* feed on the livestock of the marsh.

BELOW: Starlet Sea-anemone.

glassworts *Salicornia* spp. can grow, but the main features of the marsh are stands of brackish grasslands and rushy pastures. At its richest the familiar components of Thrift *Armeria maritima* and Sea Aster *Aster tripolium* are supplemented with Bulbous Foxtail *Alopecurus bulbosus*, Parsley Water-dropwort *Oenanthe lachenalii* and Dotted Sedge *Carex punctata*; there are also surprise elements such as a stand of Royal Fern *Osmunda regalis* which tolerates the occasional salt bath when the wall is overtopped.

All of the marsh is flooded during extreme high tides but this does not appear to have an adverse effect on seasonally parched grasslands which have formed over the more gravelly parts of the site. Here a range of coastal clovers grows, amongst the fine grasses, including Subterranean Clover *Trifolium subterraneum*, Knotted Clover *Trifolium striatum* and Bird's-foot Clover *Trifolium ornithopodioides*. Bare patches provide niches for the germination of the annual Upright Chickweed *Moenchia erecta*, a frequent enough plant of the nearby New Forest heaths but quite a rarity beyond that stronghold.

It is thought that it is the presence of parched grasslands on the marsh and seawall that supports one of the last populations of Wall butterflies *Lasiommata megera* in this part of England. Over recent years the distribution of this butterfly has retreated to the coast. The reason for such a rapid decline of a formerly widespread species has yet to be fully understood. What appears to be happening is that their preferred habitats of thin grasslands are becoming too luxuriant due to agricultural abandonment of less productive grasslands together with an overall increase in fertility across the landscape. The collective stresses of shoreline habitats counteract the effects of fertility and so sustain the skeletal grasslands required by the Wall.

Whilst Wall butterflies are retreating to Keyhaven there are other invertebrates of the marsh that are undergoing a phase of expansion. In the 1980s the Keyhaven marshes were of national importance for their populations of bush-crickets including both species of coneheads *Conocephalus* spp., together with Roesel's Bush-cricket *Metrioptera roeselii*. I can recall knowledge of the location of Roesel's Bush-cricket being treated with the same reverence as was required for the rarest orchid. In the early 1990s coneheads were regarded by Paul Sterry (1991) as distinctly southern and coastal in distribution, with Roesel's believed to be a specialist of grazing marshes. Following the hot summers of 1989 and 1990 a growing number of records started to be received of bush-crickets outside their established range. Many of these records came from inland locations where it appeared they were colonising the verges of trunk roads and onwards into otherwise unremarkable grassy habitats. Over the last 30 years bush-crickets have moved out from their saltmarsh refuges to occupy much of southern England, with each year bringing in records from further north and west. With hindsight, the bush-crickets were not specialists, they did require a certain structure of grassland but it need not be salty. The limiting factor appears to have been climatic. Today Roesel's Bush-cricket is one of the most commonly recorded members of the grasshopper family, its former rarity still encouraging entomologists to track its spread.

To many, the saltmarshes between Lymington and Keyhaven are synonymous with birdwatching. A seawall footpath overlooks the former saltworks and gives excellent views across the neighbouring intertidal to the Isle of Wight. The birdlife is what one may expect in a diverse coastal wetland, most species being associated with the structure and location of the habitat rather than a dependence on it being a saltmarsh. Bird numbers are excellent in most winters, with

Wall butterflies (left) are contracting their range into saltmarsh grasslands, whilst bush-crickets (above) are expanding into other habitats.

the embanked marshes providing high-water roosts for waders and bespoke grazing swards for waterfowl.

Lymington's intertidal marshes support ground-nesting birds including a gullery. Black-headed Gulls *Chroicocephalus ridibundus* nest directly on the saltmarsh as well as the shell banks aligned along the marsh frontages. This gullery dates from the 1920s since when the marsh has provided the birds with a clear view of predators and relative freedom from disturbance. The growth of Lymington's gullery occurred as *Spartina* marshes extended both the area and elevation of the marsh during a period when the gull's heathland nesting sites were being destroyed through forestry plantations and urban growth. Having peaked at around 8,000 pairs in the late 1960s the gullery is in long-term decline as each season progressively high tides wash out the gulls' first-laid clutches. After nearly a century nesting on saltmarshes the gulls are moving on to more reliable territories. A small but growing breeding population has now established itself on a reservoir nature reserve in the nearby Avon Valley.

Since the 1980s Mediterranean Gulls *Larus melanocephalus* have become progressively well established as a British breeding bird. Some of the earliest pairs to succeed were attracted to the gulleries of the

A mixed flock of Mediterranean and Black-headed Gulls.

Lymington River, the incomers taking up nest sites in the tall Sea-purslane *Atriplex portulacoides* of creek edges. Numbers fluctuate year on year, the peak of over forty pairs not being repeated in recent years. Unlike their more abundant neighbours, Mediterranean Gulls do not re-lay should their nests be flooded out or predated. Whilst this early foothold for the species in Britain remains vulnerable, their progeny are spreading into the saltmarshes of southern England in ever-increasing numbers. In the short term there is considerable capacity for Mediterranean Gulls to expand their population on the Lymington River estuary. In the medium term the continued shrinkage of the marsh will remove their habitat and, like their black-headed relatives, they will need to move elsewhere.

Brownfield sites are priorities for redevelopment; it takes foresight to see their potential for wildlife. It is to the great credit of Hampshire County Council that in the early 1970s they saw and secured those possibilities. Through a series of strategic purchases the former saltworks have been safeguarded as a nature reserve. The County Council's land within the seawall is contiguous with the Wildlife Trust's nature reserves in the intertidal. Working together, the statutory and voluntary sector have brought the estuaries of the Lymington and Keyhaven rivers under conservation management. The character of the landscapes derived from the former saltworks is of such exceptional quality that the coastline was included in the New Forest National Park when its boundary was established in 2005.

Nature has reclaimed the industrial coast of Lymington but at a cost. Hyper-saline conditions no longer persist in its wetlands and as the salt industry faded so did the habitats of Brine Shrimps; the creature has been added to the list of British extinctions.

Southampton's Spartinas

The emergence of cord-grasses *Spartina* spp. from Southampton Water in the 19th century is arguably the natural feature of British saltmarshes that has the greatest claim to global significance. The evolution of a new species is fascinating and much has been written about its genetics and progenitors. It is not the intention here to repeat in detail those technical elements of the story but to explore the myths that have grown up around those species, together with recognising the naturalists who contributed to their discovery.

In exploring Southampton's *Spartina*s it is necessary to use both precision and ambiguity. Amongst botanists the terms cord-grass and *Spartina* are interchangeable, the scientific name being adopted as a ready synonym for the English. Southampton's cord-grasses arose from the hybridisation of Smooth Cord-grass *Spartina alterniflora* and Small Cord-grass *Spartina maritima*. At first their progeny was thought to be a new species *Spartina townsendi* Townsend's Cord-grass. It was not until much later that it was realised that two taxa had arisen from the hybridisation. The sterile form was renamed *Spartina* × *townsendii*, recognising its hybrid origin and retaining the English name whilst gaining a new spelling. The fertile form was acknowledged as a species in its own right and named Common Cord-grass *Spartina anglica*.

In 1907 Lord Montagu of Beaulieu promoted the virtues of Southampton Water's vigorous new *Spartina* to his peers on the Royal Commission on Coast Protection. Montagu had overseen the introduction of the species to his estate in the Western Solent and had been delighted with the results. Initial enthusiasm for the

OPPOSITE PAGE:
Common Cord-grass.

Common Cord-grass
displaying well-developed
anthers.

cord-grass bordered on the evangelical. The novel grass had a remarkable capacity to accumulate sediments, so raising the surface of the intertidal by up to 2m. With minimal investment it was possible to take intertidal mud and, with astonishing rapidity, convert it into land ready for farming or development. The Beaulieu introductions of 1898 inspired plantations in North Norfolk (1902) and the Wash (1910), then through the 1920s and 1930s across England and into southern Scotland. The vigour of the cord-grasses was such that from the late 1880s they rapidly colonised the estuaries of England and Wales, by the mid-1960s dominating about a quarter of Britain's saltmarshes. Overseas territories were also to benefit from the potent *Spartina*; introductions failed in the tropics but established in the milder climes of South Australia and New Zealand. Beyond the empire the Pacific coast of the United States was colonised, as was continental Europe. The natural spread into Europe was reinforced by English material raised in *Spartina* gardens and shipped to the Netherlands and Germany. *Spartina* also spread throughout France's Atlantic coast, the English source material supplemented by the simultaneous emergence of the same hybrid on the borders of France and Spain.

It is difficult to comprehend the scale of the consequences of the spread of hybrid cord-grasses on the world's coastal wetlands. In 1963 hundreds of plants were exported to China with the intent to prepare mudflats for development. Most of the young plants died but the 21 individuals that survived in Jiangsu on the Yellow Sea went on to colonise more than 36,000ha of wetlands in only 20 years. The potential for eco-engineering was well established by 1979 when freshly imported material was introduced to the coast of Fujian Province. An *et al.* (2007) tell how, after an initially slow establishment phase, the pace of colonisation accelerated and covered more than 112,000ha by 2000.

Across the temperate world natural habitats and native species have succumbed to the spread of Southampton's *Spartina*. Fresh territories are still being colonised; as recently as 2003 *Spartina anglica* arrived by unknown means at Robert's Banks near Vancouver and only time will tell how far it may spread through the estuaries of Canada.

Evidence, myths and extrapolations

There is currently a fashion in British botany to assign plants to various categories that indicate when they arrived on our shores. The terms archaeophyte and neophyte are common currency in botanical literature. Neophytes are species not naturally native in Britain but which have become established in the wild since 1500. Archaeophytes are similar species but became established before 1500. Many plants growing in Britain have been allocated to one or other of these categories, all too often without the benefit of corroborative evidence or peer review. It is a fashion and will pass; however, the categories are useful, even if sometimes misused.

Cope & Gray (2009) present the assertions of some botanists who regard all species of *Spartina* as alien to Britain, including the Small Cord-grass. Such a claim is difficult to understand as early botanical accounts from Thomas Johnson in 1632 and Christopher Merrett in 1666 record it from credible natural locations. Most persuasive are the physical remains of Small Cord-grass described by Hillman (1981) as being retrieved from beneath a Bronze Age boat on the Humber estuary.

In Britain, Small Cord-grass is now so rare as to be considered endangered. Where it persists in any quantity, such as on the Solent saltmarshes at Newtown, it is an occasional sub-dominant component of hollows in the upper marsh amongst sea-lavenders *Limonium* spp. and Sea Arrowgrass *Triglochin maritima*. Being small, it seems to do particularly well where the shore is lightly trampled by cattle or people as it is vulnerable to being displaced by coarser

Causeway Lake, Newtown, retains a population of Small Cord-grass.

species. In his description of Small Cord-grass on the Isle of Wight the Reverend Bromfield (1856) uncharacteristically departed from his usual scientific detachment to report:

This rank-smelling grass is quite destitute of beauty; nor does it recommend itself to any known use, unless by its creeping and fibrous roots serving to consolidate the soft fluctuating soil on which it grows, and affording a safe, if not a dry, footing over the dreary waste of flat salt-marsh.

In Europe, Small Cord-grass is distributed along the shores of the Atlantic from the Netherlands to Spain with outliers in the Adriatic. Small by name and small in stature, *Spartina maritima* rarely exceeds shin height. On the opposite side of the ocean the much larger Smooth Cord-grass *Spartina alterniflora* is native to the temperate latitudes of the Atlantic seaboard of North America. It is the most common of the suite of American coastal cord-grasses where it forms dense chest-high meadows in the intertidal.

To what degree there is a natural exchange of species across the Atlantic is a recurrent theme in the study of saltmarshes. Long before

the advent of transatlantic trade, natural processes have carried plant material eastwards to establish outliers of several American species along Europe's coast. It would be unwise to deduce too much from the type specimen of Smooth Cord-grass coming not from the Americas but from the estuary of L'Adour at the foothills of the French Pyrenees. On 20 June 1803, as war reignited between Britain and France, Loiseleur-Deslongchamps records that the herbarium specimens that define the species were gathered from the marshes along the coast road to Bayonne.

Herbarium specimens of the parents of the hybrid, *Spartina alterniflora* (left) and *Spartina maritima* (right).

It is debatable when Smooth Cord-grass was first recognised in Britain. Reverend Bromfield published a comment that he discovered it growing abundantly in Southampton above the Itchen Ferry in 1836. The precise chronology is unclear but Colin Tubbs (1999) suggests Bromfield may have heard about the plant from W Baxter who in turn had been informed by J W Weaver. There is however an earlier specimen of the grass, lodged in the herbarium at Kew, which was collected by William Borrer in 1829. In his 1904 *Flora of Hampshire* Frederick Townsend accepted Bromfield's claim for making the first record and observed 'A labourer informed Dr Bromfield in 1836 that he had known the plant at Southton for upwards of twenty years. It

The Reverend William Arnold Bromfield.

Smooth Cord-grass,
Hythe, 2016.

might have been introduced from America'. What Townsend was actually referring to was an editorial footnote to Bromfield's account by Sir William Hooker who had also visited the site and had spoken to a labourer. Bromfield's original record reported contradictory discussions with various labourers including an old man who had never known a time when the grass wasn't there. Whilst local people may not have appreciated its scientific interest, Bromfield reported that the grass was valued and cropped by 'the poorer classes of Southampton' for animal bedding and for thatching outbuildings. He concluded his account by remarking that every accessible stand of the grass was cut in September as 'a plant of real economic utility'. In a more egalitarian time perhaps the first record would have been credited to those unknown labourers.

Among his numerous footnotes to Bromfield's account, Sir William was open to considering Smooth Cord-grass to be 'an aboriginal native of Europe'. Townsend's later emphasis that it 'might have been' introduced has grown over time into unqualified assertions that it was certainly imported from North America, if not deliberately then accidentally. In the 19th century Southampton was an important port and remains so today. Various authors have speculated on the introduction of Smooth Cord-grass through its survival in ballast or bilge water, as dunnage around fragile goods, as hay for ship-borne livestock and even mislaid fibre imported for making banknotes.

No author has offered evidence that their chosen vector was actually the case. The means of introduction, so boldly asserted, are little more than conjectural.

Whilst the origins of Smooth Cord-grass are obscure, there is clear evidence its populations were supplemented by at least one introduction. Specimens in the herbarium of the Natural History Museum in London are accompanied by a letter of 1956 describing a glabrous form of the grass being imported by Professor Oliver in 1925 which he planted into Poole Harbour and Eling, the later site being at the head of Southampton Water. The letter goes on to report that the Poole introduction had failed and the Eling population had declined to just one tussock.

The Hampshire coast was something of a botanical backwater before the mid-19th century. Records are sparse and so we cannot be sure there was a time when Smooth Cord-grass was absent. The first records as collated by Townsend (1904) demonstrate that many of the commoner saltmarsh species are contemporary with, and some post-date, the 1829 record for Smooth Cord-grass. Small Cord-grass had been known from the Solent since *c*.1770 but the first Hampshire record for Sea Aster *Aster tripolium* was 1835, followed by Fat-hen *Chenopodium album* and Annual Sea-blite *Suaeda maritima* in 1837. Sea-purslane *Atriplex portulacoides* records date from 1841, with Saltmarsh Rush *Juncus gerardii* and Sea Club-rush *Bolboschoenus maritimus* both in 1844. At the time that botanists were systematically working the Solent, the Smooth Cord-grass was a well-established, if rather localised, component of its saltmarshes.

Hybridisation

Evidence of the origins of Smooth Cord-grass is insufficiently substantial to prove an introduction or to dismiss the possibility of a natural outlier of the American population. What is without question is that at some unknown time, possibly in the first half of the 19th century, the two species of cord-grass, the Small and the Smooth, hybridised.

Anyone who has spent time looking at *Spartina* stands will appreciate that descriptions in literature don't always help define material in the field. In my student days the standard reference book was *Grasses* authored by Charles Hubbard, a world authority on the classification of grasses based at Kew. It was Charles Hubbard who recognised that the fertile form of the hybrid cord-grasses was actually a newly evolved

species and who published the definitive description of *Spartina anglica* in 1978.

Understanding what had happened to Townsend's Cord-grass, the first progeny of the hybridisation, was greatly assisted by the development of techniques to investigate the cytology of grasses. Studies by Chris Marchant and colleagues at the University of Southampton in the 1950s and 1960s proved the answer to be genetic. Like many hybrids, Townsend's Cord-grass was sterile, incapable of setting viable seeds or fertilising another plant. At some unknown time and place Townsend's Cord-grass had gone through the process of allopolyploidy. In simple terms some chance event resulted in a doubling of its chromosomes through which the grass gained fertility with the theoretical possibility to produce plants true to itself. The hybrid had evolved into a fully formed species, albeit one that appears to be self-incompatible when it comes to setting seed.

Hubbard's *Grasses* was originally published in 1954 with many reprints, as well as new editions in 1967 and 1992. By 1967 Charles Hubbard had sufficient confidence to include a description of Common Cord-grass in *Grasses* even though it had yet to be published in the scientific press. The last revision of *Grasses* was edited by his son, John, who maintained the family interest in all things graminoid after his father's death in 1980. John took a particular interest in the ecology of *Spartina* and contributed extensively to the scientific literature. I therefore eagerly accepted an invitation to join him when in 1994 he wanted to visit a nature reserve I helped care for at Hythe. John and I wandered the marsh, gathering specimens and pondering their variety. It was a deeply reassuring walk. Not everything could be named and I learned from Hubbard himself that it was acceptable not to know; *Spartina*s really are that difficult.

The Hythe *Spartina* Marsh nature reserve has a special place in the *Spartina* story as it is supposedly the place on Southampton Water where the hybrid *Spartina* was first collected. Whether this was the actual site is open to question, as a detailed account by Frederick Rayner (1907) of the Solent's *Spartina* referred to the marshes immediately to the south of Hythe pier being the *locus classicus*. Hythe grew substantially over the 20th century; the intertidal close to the shore has been built over so that today there are no marshes within a kilometre of the pier. Much has changed around the *Spartina* Marsh nature reserve since the early 20th century; not least the face of the marsh has contracted over 200m. What was a deeply rural coastline is now backed by housing estates and an oil refinery. Probably the most ecologically significant change to the habitat came with the fencing of the estuarine commons along the east of the New Forest in the late 1960s. To keep commoners' livestock safe from traffic it was deemed necessary to confine them to the west of the trunk road. The evolution of cord-grasses at Hythe, and along much of Southampton Water, progressed under the stresses imposed by free-ranging ponies and cattle. With their removal the continuous grazing and trampling pressures on the habitats have been greatly curtailed to the disadvantage of all but the most robust species.

James and Henry Groves

Credit for the recognition that the Smooth Cord-grass was not the only large *Spartina* in Southampton Water goes to the Groves brothers, James and Henry.

When Henry died in 1912 his younger brother wrote his obituary for the *Journal of Botany* (1913) in which he gave an affectionate summary of an extraordinary botanical career. The Groves family were respectable but of limited means, so when his father died Henry had to leave school and make a living; he was only 14. His school,

Henry Groves.

Godalming Grammar, was set in the Surrey Hills with heath and down offering fertile ground for inquiring minds. The headmaster, Mr Churton, was a keen naturalist and tutored favoured boys in the skills of observation and critical thinking. With limited schooling Henry had none of the opportunities of a 'varsity man; instead he had the drive for self-improvement and a brother to share his passion for natural history. James thought they took life rather too seriously with little time for amusements. It is difficult to reconcile this view with the evident manner in which they made themselves agreeable to others. Despite not having the private incomes or public positions of their intellectual peers, the brothers built a network of friends amongst the most eminent botanists of their day.

The natural history skills of James and Henry were formidable. Before the age of 20 Henry graduated from commonplace botany to embrace the challenges of critical species. His interests were not just botanical but extended to mud snails and algae. Henry and James worked together on the classification of stoneworts, including the definitive description of the Foxtail Stonewort *Lamprothamnium papulosum*, a great rarity of saline lagoons. Their unravelling of the relationships between species in this highly complex group of algae remains a standard reference in their study today.

Between 1873 and 1879 summer holidays were taken together around Southampton. The brothers had become correspondents and co-workers with the older Frederick Townsend, a gentleman of substantial means who lived in south Hampshire. Townsend shared their passion for critical botany and was working on particularly demanding groups including glassworts *Salicornia* spp. and eyebrights *Euphrasia* spp. It was he who determined that the minute eyebright *Euphrasia foulaensis* of saltmarsh turf in the Highlands was a species new to science.

The brothers' interest in Hythe was possibly stimulated by reports that Robert Southerby Hill, a doctor from Basingstoke, had found an unusual *Spartina* there a few years earlier. David Allen (1996) tells that Hill was highly regarded for his knowledge but had 'an insuperable aversion to publishing his opinions or observations' and so gained little recognition outside his immediate circle.

The critical faculties of the brothers were applied to the grass from Hythe; it was distinctly different from Smooth Cord-grass and it was definitely not a luxuriant form of Small Cord-grass as suspected by Dr Hill. It took four years between gathering the specimens and publishing

their account (1882) of a species new to science; they chose to call it
Spartina townsendi in recognition of the friend who had encouraged
their endeavours. Men in their early twenties of no status or academic
background were not expected to do such things. James recalled:

> *I well remember the late Reverend W.W. Newbould shivering at our*
> *audacity, evidently thinking it scarcely decent for young men to describe a*
> *new British plant without some first rate botanist like the late William*
> *Borrer having viewed it* in situ *to see that it was all right!*

It clearly was alright. The year 1878 had been productive for the
Groves at Hythe; not only had they collected a species new to science
but Henry found another new to Britain. He discovered Adder's-tongue
Spearwort *Ranunculus ophioglossifolius* growing in a lane-side pond where
the New Forest's heaths run down to Hythe's saltmarsh shore.

The Groves brothers shared the enthusiasm of the botanists of their
age in being avid collectors of the rare and novel. Dried specimens
of *Spartina* from around the Solent fill many cupboards in Britain's
herbaria. In the British Museum there is a sheet labelled *Spartina
alterniflora* collected by George Edgar Dennes in September 1846
from the upper reaches of Southampton Water. The label gives the
precise location, Itchen Ferry. Dennes was a solicitor who served as
the secretary to the Botanical Society of London. He laboured in that
role alongside the overbearing intellect of Hewett Cottrell Watson.
David Allen's history of the Society (1986) suggests that Watson was
the driving force behind the Society and had the greatest respect for
Dennes' exertions but little regard for his botanical skills. Contrary
to Watson's judgement, Dennes' abilities were substantial as he had
an eye for the unusual; Manchester Museum holds specimens of the
hybrid Perennial Beard-grass × *Agropogon littoralis* that he gathered
from the Thames marshes. It is probable that he recognised that the
cord-grass at Itchen Ferry did not conform to the norm, but he was
not in a position to offer an alternative name.

There are no saltmarshes today at Itchen Ferry as Southampton has
grown over them. When Dennes visited there were extensive marshes
on the edge of the city and those by the ferry were a convenient place
to botanise as they lay by the main road to Portsmouth. Other botanists
were attracted to this accessible spot; in 1850 Frederick Townsend
recorded Smooth Cord-grass as being profuse on the tidal Itchen, with
Joseph Wood's notes on Small Cord-grass giving the precise location as

the Itchen Ferry. By these records we know both species were present in the same place at the same time.

In 2011 Dennes' herbarium sheet of *Spartina* was reconsidered by Tom Cope and Mark Spencer. Tom Cope had succeeded Charles Hubbard as the national referee for grasses and was reviewing the material curated by Mark Spencer. The specimen has been carefully prepared by Dennes with its critical features preserved intact; Spencer and Cope re-determined the material as Townsend's Cord-grass *Spartina × townsendii*. The Groves may have published their discovery in 1882 but Dennes provides proof that the hybrid existed nearly four decades earlier. The compact chronology of the first records of the hybrid and its parents at Itchen Ferry is to a degree a matter of chance. All three cord-grasses were present when botanical investigations reached the Solent's backwaters; it begs the question as to how long they had already been there.

Expansion and contraction

There is circumstantial evidence that *Spartina* may have been responsible for a small expansion of saltmarshes in the upper reaches of Southampton Water in the early 19th century. By the end of the century that relationship was beyond doubt. By the late 1880s stands of large cord-grasses, by now named *townsendi*, were springing up around the Solent with no obvious connections to the original population. Townsend's Cord-grass proved to be a vigorous colonist but as it was sterile it could only spread by vegetative means. Unquestionably, some of the new populations were due to deliberate introductions, as at Beaulieu, but not all. A suggestion as to what was happening came in 1892 when a fertile form of Townsend's Cord-grass was recorded from the Lymington River. Unlike the empty anthers of the original *townsendi* the new form bore healthy pollen. It has been supposed that with fertility this *Spartina* had a second means of dispersal; if viable seed could be set then it may be distributed by the tide. Theory and practice differ; *Spartina* seedlings are unrecorded in the wild. The pollen-bearing *townsendi* appears to have had greater vigour than the original form and so natural dispersal of root fragments was probably responsible for the spread of *Spartina* accelerating throughout the Solent and beyond. The fertile form rapidly became recognised as the most widespread of all the cord-grasses, duly earning its vernacular name of Common Cord-grass.

In retirement James Groves left London and moved to the Isle of Wight. He never lost interest in natural history and kept a close eye on the *Spartina* he and his brother had discovered in their youth. The venerable naturalist rejoiced in his *Spartina*, setting aside the devastating consequences of its spread on the other life of the marsh. Shortly before his seventieth birthday Groves (1927) addressed the Isle of Wight Natural History and Archaeological Society:

> *Perhaps not many of those present can remember the entrance of Lymington River fifty years ago. I have a vivid recollection of how one used to go out of the river at low tide between horrible-looking and malodorous banks of bare, half-liquid mud. Thanks to our Spartina the state of things now is quite different, and it would be difficult to find a more beautiful sight than that of the Spartina-area in the same spot on an autumn evening with the setting sun shining on the golden stems and leaves.*

Before Groves died in 1933 the tide was already on the turn for his cord-grass. The first indications of loss of vigour of Southampton's *Spartina*s came from Poole Harbour, a little to the west of the Solent. The first record of a large *Spartina* in the harbour was from 1899, a single plant that over the next 20 years went on to dominate nearly 900ha of the intertidal. By the early 1920s the pattern of colonisation was becoming more complicated; tidal flats were still being colonised

James Groves' grass dominating the mouth of the Lymington River.

but some of the older stands had started to collapse. By the end of the decade losses were outpacing colonisation and the marsh had started to recede.

In the late 1920s similar reports were being made from the Beaulieu River where the *Spartina* stands promoted by Lord Montagu to the Royal Commission were starting to break up. It was not long before die-back was being reported from across the Solent. On the seaward face of the marsh the plants most exposed to wave action were being scoured away whilst within the creeks and pans their rhizomes were rotting. Infections were detected in the rhizomes; however, these didn't appear to have been the cause of death but symptoms of stress from another source. What was weakening the *Spartina* stands was that they had outgrown their habitat. The super-efficiency of the grasses in accumulating silt had led to stagnant conditions; fine silts trapped by the grasses were sealing off the marsh surface from essential root aeration. The anaerobic conditions resulted in concentrations of compounds of iron and sulphur that reacted to further reduce oxygen levels and so the weakened marsh collapsed, tidal scour redistributing the sediments back to a proximate of former levels.

It is almost impossible to summarise the consequences of the expansion and contraction of *Spartina* stands on the Solent's wildlife as everything was affected. Species dependent on the open shore were squeezed out as *Spartina* built its mudbanks and marshes. James Groves may have considered mudflats malodorous but those alongside the Lymington River once supported eelgrass *Zostera* spp. beds with rich foraging grounds for wintering birds, an abundance of shellfish and nursery grounds for Bass *Dicentrarchus labrax* and other commercially important fish. There are signs that the new alignment of sediments may recolonise with something broadly similar. In the meantime, circumstances have changed so that the ecosystem never goes back to where it once was.

The current status of *Spartina* in the Solent

The pulse of changes driven by the rise and retreat of Southampton's *Spartina* is far from over. Since saltmarsh species were first recorded in the Solent the fortunes of *Spartina* have changed so it is timely to take stock of the current situation.

Small Cord-grass has declined since being described by Townsend as rather common, to its current status of exceptional rarity. The

grass persists on the Isle of Wight in the sheltered upper marshes of Newtown Harbour, those marshes grading into gently farmed grasslands and ancient woods. On the mainland the last records from the Western Solent were made in the 1970s. In the east of the Solent a few plants persist on North Common on Hayling Island and so it has not yet been lost from the mainland shores. The decline of Small Cord-grass can be partially attributed to being displaced by its more robust relatives, an ecological succession that has been exacerbated by agricultural abandonment and subsequent dereliction of upper marsh communities. As well as loss of quality, the saltmarshes of the Solent are being squeezed out by the sea, a combination of natural landforms, flood defences and urban growth curtailing natural landward migration of the high tideline.

Smooth Cord-grass survives in Southampton Water but in much reduced circumstances. In 2015 Martin Rand and I spent an evening at Bury Marsh, a few kilometres from Hythe, as this was reputedly the last wild stand in the country. Despite having access to detailed records from previous surveys we were unable to find any material we could assign to *alterniflora*. We were not so bold as to seek to declare it absent on the Bury shore; what we concluded is that we searched in vain.

Bury Marsh.

Specimens of *Spartina × townsendii* dwarfed by *Spartina alterniflora*, Hythe.

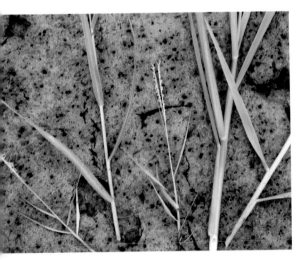

There is a stand of *Spartina alterniflora* growing on the landward edge of the Hythe *Spartina* Marsh, just outside the nature reserve fence. It appears to be a clonal patch and so presumably derived from a single point of establishment. Jack Coughlan, who has known the marshes since the 1960s, advised me that he had been told the colony had been planted. By whom, when and why such an introduction may have been made is unrecorded and Jack recognises this may just be another myth embellishing the *Spartina* story. Evidence in favour of this population being an introduction is that it is conveniently sited near the road and has grown out from a sub-optimal habitat, the person making the introduction possibly being unaware of its preferred habitat at the face of a marsh. The Hythe *alterniflora* appears healthy and seems to be spreading. Rand & Mundell (2011) report its status in 2006 as a patch 6×5m with a few isolated plants beyond the margins; in 2016 it was dominant over an area of some 18×20m.

Separating *Spartina × townsendii* from *Spartina anglica* in the field is not as straightforward as the books imply. From current British records it appears × *townsendii* is in steep decline, having somehow lost its initial hybrid vigour. The difficulty of identifying specimens in Southampton Water with confidence is a reflection of the variation in what is growing there. It has been suggested that further hybridisation or genetic drift may be occurring with the possibility of the emergence of one or more additional forms. The hybrid × *townsendii* certainly persists at Hythe and Bury but is becoming increasingly rare elsewhere.

Common Cord-grass remains common across the Solent's estuaries despite the process of die-back that began in the 1920s still being active today. In the sheltered estuary of Langstone Harbour at the eastern end of the Solent the process is almost complete, with mudflats restoring themselves to something akin to their former status. In Langstone *Spartina* is being replaced by eelgrasses *Zostera* spp., a group of species whose populations had collapsed in the early 20th century due to a wasting disease. On the open shore of the Western Solent the face of the marsh recedes as the creeks broaden and flatten behind them. The pace of change is tangible even to the unobservant. What was a sheltered shore is progressively exposed to the reach of wind and tide.

Langstone Harbour, Hampshire, where die-back of Common Cord-grass is almost complete.

Saltmeadow Cord-grass,
Chichester Harbour,
West Sussex.

At the foot of the eroding marshes a lower alignment of sediments is becoming available for colonisation. Every so often a clump of Common Cord-grass becomes established on these lower shores; presumably, the refreshed sediments are better aerated than the high marshes. Where I have seen this happening these isolated clumps appear to be stable and have yet to demonstrate the capacity of their forebears for exponential colonisation. After its unsettled adolescence this newly formed species may have found a place in the marsh communities of Southampton Water.

All good tales of jeopardy, ecological or otherwise, should conclude with the hint of a fresh peril. There is a new *Spartina* in the Solent. In 2013 a brief report by Geoff Hounsome was published in *BSBI News* confirming the discovery of Saltmeadow Cord-grass *Spartina patens* in Chichester Harbour, not far from where Charles Hubbard collected the type specimen of *Spartina anglica*. Tony Spiers' preliminary identification of 2005 had been confirmed beyond doubt by both Eric Clement and Filip Verloove. Saltmeadow Cord-grass grows alongside Smooth Cord-grass on the Atlantic coast of North America. It has a broad range of tolerances assisting its growth through upper saltmarshes and into freshwater wetlands.

It is necessary to consider whether *Spartina patens* could become an economically important invasive non-native species in Britain. In 2012 there was just a single patch by the seawall in Chichester Harbour, a mere 8×3m, which had grown to 14×5m by 2016. From what we know of its habitats in North America, the British population is in a sub-optimal position in the estuary in not having access to the brackish reaches of upper saltmarshes. The consensus at present is that it is necessary to monitor the situation but there is no urgent case for extermination.

Given that *Spartina anglica* and × *townsendii* arose from hybridisation, it is worth considering whether *Spartina patens* could contribute to the formation of novel hybrids; evidence from North America suggests not. *Spartina patens* is one of the fine-leaved *Spartina*s, in the same genus but distinctly different from members of the family known from Britain. Hybrids of *patens* are unknown in the wild.

The closest relative to *Spartina patens* in Europe is *Spartina versicolor* which has long been regarded as native to the Azores and Mediterranean coast. *Spartina versicolor* is a widespread if under-recorded species; when it is not in bloom it may be overlooked as just another fine grass of the shore. The recent discovery of a fine-leaved *Spartina* reported by Izco *et al.* (1999) on the Atlantic coast of Galicia has raised the question as to whether *patens* and *versicolor* are different species; they are very difficult to tell apart using their physical characteristics and genetically they are similar. The European *versicolor* appears to be a form of *patens* but with a lesser genetic diversity, suggesting *versicolor* emerged from *patens* through an evolutionary bottleneck sometime in the unknown past. That bottleneck could have been caused by the current populations establishing from very few plants, possibly even just a single specimen. That original colonist probably arrived from the Americas and could have done so in this, or previous, interglacials.

The source of the fine-leaved *Spartina* in Chichester Harbour is a matter for conjecture. Only genetic analysis can determine whether the material is the same as the *versicolor* of Europe or the *patens* of America. The current best judgement is that it is of American origin, possibly by way of the horticultural trade as *patens* has been offered in recent years by at least one British nursery.

The *Spartina* story has yet to conclude. Separating the myths from the evidence will help to bring clarity. Even in a country with a long tradition in the study of natural history it is impossible to say when and where globally important events occurred. In Britain the emergence of the new cord-grasses is at best a curiosity, at worst a local challenge. We even have the luxury of debating the philosophical niceties as to whether the species may be truly native. On the world stage the spread of Southampton's *Spartinas* remains a significant threat to the wildlife of the planet's wetlands and so introduces the issue of why and how people have sought to conserve saltmarshes.

Conservation before conservationists

S altmarshes are highly valued landscapes and have long been the subject of intense competition for access to their natural resources. The earliest exponents of saltmarsh conservation were medieval peasants motivated by economic security. Having rights to exploit a marsh underpinned the prosperity and independence of many coastal communities. Attempts to extinguish those rights have been fiercely resisted.

Unknown sons of iniquity

Accounts of the earliest resistance to the destruction of saltmarshes have survived in the records of the destroyers. Rights enjoyed by the peasantry who were governed under the English manorial system were established by custom and practice as the written word was of little use to a predominantly illiterate society. Fortunately, there are collections of documents that provide accounts of saltmarsh management in the medieval period. These records are inevitably partial, particularly when it comes to financial or legal affairs. With major religious houses holding estates in the Somerset Levels there is a particularly rich source of manuscript records from those wetlands.

The estates of Glastonbury Abbey date from the 8th and 9th centuries. Within the Abbey's estates there were commons derived from saltmarshes, modified by rudimentary sea defences but still vulnerable to the vagaries of tides. Traditional rights varied but usually included the ability of commoners to graze livestock and in some cases extended to taking wildfowl. A series of legal disputes in the mid-13th century on the Huntspill Levels arose following resistance to the abbot's

OPPOSITE PAGE:
An abundance of life
on the Wash.

attempts to enclose such a common. The peasants who risked losing their livelihood resorted to direct action, breaking down seawalls and diverting the watercourses. When miscreants were apprehended the court records indicate that the marshes were not only of importance to the local community but also to rights-holders from further afield. These prosecutions indicate that livestock were driven down from the Polden Hills to the summer pastures of the coastal levels. On the opposite shore of the Severn in the Gwent Levels this tradition of transhumance survived enclosure; in 1954 the Welsh Agricultural Land Sub-Commission found 8 per cent of the holdings on the Levels were summer-grazed pastures linked to farm enterprises high in the adjoining hills.

Direct action against the ecclesiastical estates inevitably failed but the defenders of the marshes sometimes evaded capture. In 1315, having failed to arrest the defiant perpetrators, a vexed Dean of Wells condemned his tormentors as 'unknown sons of iniquity'.

Resistance and riches

The power of the ecclesiastical estates may have been broken by King Henry's reformation but this did nothing to diminish the influence of the Crown. In the early 17th century King James I of England and his successor Charles were both attracted by opportunities for income from the Crown estate. Having established a legal precedent that land formed by the seaward growth of a marsh belonged to the Crown, they set about to realise those assets. The land created by expanding marshes may have belonged to the Crown but it was still subject to the rights of others. The King did not have a free hand to do as he pleased. One option was to sell the land to those who already had the ability to exploit it. David Robinson's studies of the Wash (1987) show that in 1613 King James sold the freehold of the Crown's portion of saltmarsh commons in Moulton and Long Sutton (Lincolnshire) to its commoners. Two years later further sales were made elsewhere in the county at Wigtoft, Whaplode, Holbeach and Tydd St Mary. The common pastures available to communities on the Lincolnshire coast could be extensive; at this time the contiguous commons of the neighbouring parishes of Moulton and Gedney extended to some 1,820ha.

Where commoners had security of tenure, together with the capital and a collective will, there are cases where saltmarshes were embanked for the purposes of improving the communal resource. An inexpensive embankment would reduce the risk of flooding without excluding the highest tides. The saltmarsh was therefore modified but not transformed into a different habitat. Such embankments would be of great convenience to commoners with livestock that otherwise needed to be moved off the marsh each high tide. An embankment reduced the number of times those animals would have to be driven to safety. On the Lincolnshire Wash, at Holbeach and Whaplode, the commoners embanked the saltmarsh with the permission of their manorial lord. Also in Lincolnshire, Rex Russell's 1968 history of the Crown manor of Barton-Upon-Humber illustrates the Humber Bank protecting *c.*443ha of saltmarsh commons from all but the severest winter floods. At Barton there was sophisticated compartmentation of the common into cow pastures, horse pastures and unspecified Ings pastures. This pattern of use survived until enclosure in the last decade of the 18th century. Commoners were not averse to embanking marshes if it was to their advantage.

The value of the Crown estates would be greatly enhanced if those common rights were extinguished. This was possible through lawful enclosure after which the land could be embanked and transformed into farmland. A series of grants was therefore given by the King to courtiers and merchant adventurers, supported by the engineering prowess of the Dutch. Through those grants the coastal commons of the Wash and adjacent Fens were planned to be taken from the will of the waters and made profitable.

There are a few insights into the wildlife of the marshes on the advent of modern drainage. Having been occupied by people for thousands of years, the wildlife of the 17th-century coast had been substantially modified from its natural state. In common with other coastal regions Wild Cattle, Aurochs *Bos primigenius*, had survived into the Bronze Age when they were eventually usurped by domestic livestock. During that same period Dalmatian Pelicans *Pelecanus crispus* ceased breeding in the estuaries of the Cambridgeshire fens; Yalden & Albarella (2009) recognised this loss as contributing to the historical contraction of the species from across northern Europe.

On the southern margins of the Wash the saltmarshes of coastal Cambridgeshire were within reach of the attentions of John Ray, one of the finest botanists of his day. In a catalogue of 1660 Ray recorded a suite of saltmarsh plants from the wetlands of the tidal Nene around Wisbech. His account includes Marsh-mallow *Althaea officinalis*, Sea Club-rush *Bolboschoenus maritimus*, Sea Spurrey *Spergularia marina*

John Ray.

Marsh-mallow can still be found by brackish fenland drains.

and Sea Aster *Aster tripolium*. Ray's fragmentary records suggest brackish marshes through the presence of Marsh-mallow and Sea Club-rush. Sea Spurrey indicates broken ground whilst Sea Aster suggests the presence of mature marshes relatively high in the tidal range.

A fuller picture of the area can be derived from the *Flora of Cambridgeshire* published in 1860 by Professor Charles Babington. The extent of saltmarsh in the county is limited so we can be reasonably confident that Babington was revisiting the marsh country known to Ray. Babington's account described the flora of Cambridgeshire's saltmarshes at a time when the steam pumping of the embanked marshes was becoming ever more efficient. Babington's lists included species which 'have not been seen by any botanist for many years … from that that was formerly saltmarsh, but is now nearly, or, for the most part, quite dry', he observed that 'the extensive salt-marshes are nearly gone'.

What Babington considered an impoverished flora contained species we regard today as of great interest due to their scarcity and as indicators of exceptional habitats. These included Sea-heath *Frankenia laevis*, a prostrate component of short gravelly marshes, Marsh Sow-thistle *Sonchus palustris*, the towering herb of brackish margins, and Sea Barley *Hordeum marinum*, a diminutive grass of broken salty places. All of these species are now extinct in Cambridgeshire.

ABOVE: The Great Ouse at King's Lynn; draining the saltmarsh country of Cambridgeshire and Norfolk.

LEFT: Sea Barley has been lost in Cambridgeshire to agricultural intensification.

Less precise in locality, but joyful in their exuberance, are the descriptions of birdlife in 1622 by Michael Drayton in *Poly-Olbion*. Book 25 of this expansive topographical poem celebrates the bounty of the wetlands of Lincolnshire. The verses are a dense blend of classical allusion, regional economy and the delights of the wild.

225

Drayton's abundance of Teal suggests extensive marshes of short grasses.

Poetically, the 'ayre is darkened' with the multitude of duck including Eurasian Wigeon *Anas penelope*, Eurasian Teal *Anas crecca* and Common Goldeneye *Bucephala clangula*, which Drayton simultaneously found wondrous and a bounty to be taken. In amongst the swamps

> *The buzzing bitter[n] sits, which through his hollow bill a sudden bellowing sends, which many times does fill the neighbouring marsh with noyse, as though a bull did roare.*

Michael Drayton.

As well as the ducks and the Bittern *Botaurus stellaris*, Drayton describes over 30 species of bird, some familiar such as Goosander *Mergus merganser*, Common Crane *Grus grus* and Osprey *Pandion haliaetus*, others obscured by now archaic language including sea-meaw, sea-pie and puet. The saline wetlands are reflected in his association of Puffins *Fratercula arctica* with 'the brack' rather than the still clear freshwater favoured by the Little Grebe (Dabchick) *Tachybaptus ruficollis*. Amongst the estuary birds are Knot *Calidris canutus*, godwits *Limosa* spp. and the trusting Dotterel *Charadrius morinellus*, which Drayton particularly enjoyed as 'a very dainty dish'.

RIGHT: 'The buzzing bitter[n] … a sudden bellowing sends'.

BELOW: Dotterel; both trusting and delicious.

The Wash retains some of the exuberance celebrated in *Poly-Olbion*.

Anthony van Dyck's self-portrait with Endymion Porter.

At the time Drayton was writing, the Lincolnshire shore of the mouth of the Humber was dominated by extensive saltmarshes, locally known as Fitties. The villages of North and South Somercotes had, as their names suggest, grown up around summer grazing and the produce of the marsh. The local economy and way of life came under threat in 1632 when King Charles granted Endymion Porter the right to embank and enclose *c*.810ha of the saltmarsh. Porter was a favourite of Charles, not least for his diplomatic skills. A linguist, married to a Roman Catholic, he served the King on missions to Spain and the Netherlands.

The royal grant to enclose was made without proper reference to the people who depended upon the Fitties for their livelihood. Porter's enclosures would no longer be available to the people by right. The essential character would change as embanking and enclosure would enable the land to be drained and ploughed. Those who could pay the rent may become tenants and adopt a farming lifestyle; everyone else would need to find another way to make a living. Enclosure redistributed the wealth of the marsh from the many to the few.

In response to the outrage of the local community, Porter offered a secure right of way to the shore through his fields for the use of fishermen and wildfowlers. He further proposed *c.*200ha of the enclosed marsh to be set aside for the benefit of the commoners. Despite continued protests, he instructed embankments to be raised and sluices installed. Shortly after completion the engineering was attacked in the first of many acts of resistance that were sustained for nearly 20 years. In 1641 Porter eventually sought the support of the House of Lords for an order for 'Quieting his possessions' as people continued to break into the enclosures 'in a tumultuous manner'. Even in the year of his death, 1650, Porter had to eject protesting villagers who defiantly occupied his farms, still refusing to acknowledge the legality of the enclosure.

ABOVE AND BELOW: The drains of Porter's Marsh are pumped to meet their seafall.

Rancorous disputes such as Porter's Marsh were not isolated and similar resistance was occurring amongst the marshes at the headwaters of the Humber. The Dutch engineer Cornelius Vermuyden had been engaged to drain the royal estates of Hatfield Chase. The Chase was a complex landscape of tidal rivers and saltmarshes grading into peat moors and raised bogs. This was one of the Crown's larger estates, though it yielded very little to the exchequer in its unenclosed state. Whilst the Crown owned the land they did not own the grazing nor many other products of the wetlands. These elements of the property were subject to common rights which were enjoyed by people living around the Isle of Axholme. The Crown knew that if these traditional rights could be extinguished then profits could be made from new farms and high rents.

The saltmarshes of the Rivers Ouse, Idle and Trent which define the Isle of Axholme were deemed particularly suitable for improvement through the process of warping. This technique involved constructing low banks and ditches on the intertidal which were aligned to delay the discharge of water on a falling tide; the impeded drainage stimulated sediments to settle. Through successive tides the land surface could be raised through the application of fertile silts and so create rich farmland. In Axholme today the former common saltmarsh bordering Crowle Waste is still known as The Warpings.

Attempts by Vermuyden and his men to drain the marshes were met with fierce resistance. The mostly foreign workforce was subjected to a campaign of harassment and haranguing. Implements were burnt and the workforce stoned. In desperation, commoners without legal remedy to the loss of their livelihoods turned to rioting. The ensuing violence resulted in fatalities, with official records naming those who died, such as Robert Coggan who was killed during two days of bloody skirmishes in 1628. At Epworth in the following year those convicted of riot had massive penalties imposed on them, with five men each being fined £1,000 and nine women £500. It was impossible for such sums of money to be found by anyone outside the nation's elite. Remission was offered should the rioters sign an agreement to the enclosure. It was increasingly apparent that if the King's aspirations could not be secured lawfully then they would be achieved through chicanery and coercion. The commoners maintained their refusal to accept the changes even after their marshes had been transformed into farmland. In 1633, at Axholme, they entered newly made farms, taking the crops and issuing threats to the servants of the Dutch.

In doing so, the commoners sought to provoke court appearances where they could argue their case that the land had been unlawfully taken from them.

There were gentry, some enjoying common rights themselves, who took the grievances to London. The House of Commons was petitioned, parliamentary committees were established and cases were heard by the Privy Council and Star Chamber. Amongst these gentry was Oliver Cromwell, Member of Parliament for Huntingdon. The actual role of Cromwell in these disputes is clouded by his subsequent achievements. Reliable contemporary accounts report him being eloquent in his speeches to the dispossessed. He certainly presented petitions from commoners to Parliament but there were also allegations he was implicated in an unlawful fund to support rioters and their legal defence. However, the title assigned to him of 'Lord of the Fens' did not derive from the protestors but was concocted much later by propagandists who sought to ridicule him.

The Humberhead rioters were concerned with preserving their commons from enclosure; they did not articulate sentiments beyond their grievances. In his history of the Fenland riots Keith Lindley (1982) suggests Cromwell and his associates were more subtle and politically minded. To them the contested attempts at enclosure were another example of the high-handed application of power by a monarch ruling by divine right. Through supporting the common people they built political capital to challenge their king. The prolonged resistance to the destruction of the Humber's saltmarshes contributed to the breakdown of the established order into the chaos of civil war and revolution.

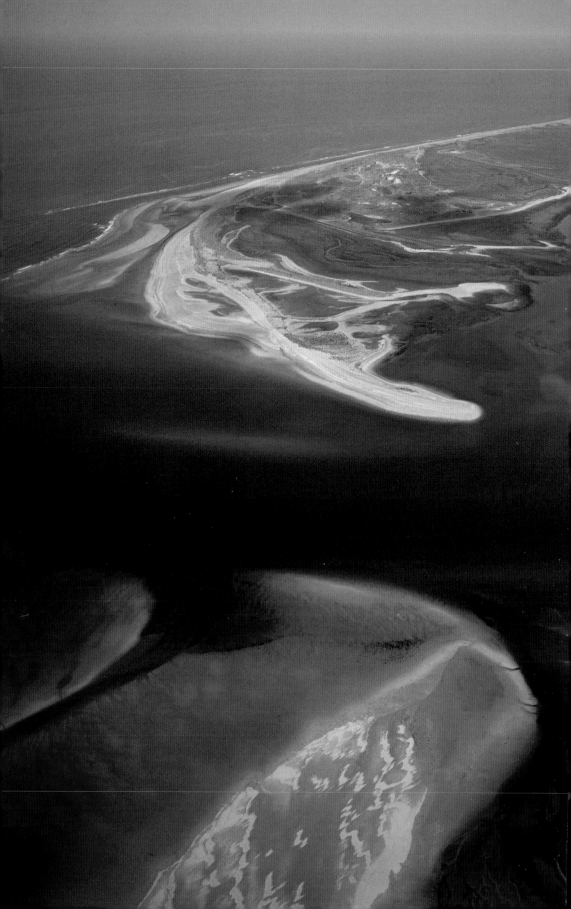

The advent of nature conservation

T he industrial revolution brought with it wealth but also dispossession and environmental degradation, and so the conservation movement grew alongside industrialism and enclosure. As the 19th century progressed so did the appreciation of wild places for their own sake, for their beauty and diversity of life. Saltmarshes were amongst the first habitats to be promoted as nature reserves, a movement that made a faltering start in the opening decades of the 20th century.

The earliest initiative to conserve a saltmarsh as a nature reserve came in 1910. Francis Oliver, an academic botanist from University College, London, purchased the old lifeboat house at Blakeney on the North Norfolk Coast to establish a base for field studies. Oliver valued the diverse habitats of the dunes and saltmarshes around Blakeney Point as a place to immerse his students in their subject. The house was also an answer to the problem that he posed in his paper of 1913 on how to make good his escape from London for at least six months of the year.

Having secured his place on the coast, Oliver became aware of its vulnerability and so he set out to persuade the newly formed National Trust to acquire the land and so prevent any interference with its ecology.

Oliver's wish was partially met when a substantial part of Blakeney Point was acquired anonymously by Charles Rothschild and fellow benefactors in 1911 and then handed over to the National Trust. His daughter, Miriam Rothschild, writing with Peter Marren in 1997, emphasised that the gift was given on condition that the natural flora and fauna should be preserved. At 445ha this was a large nature

OPPOSITE PAGE:
Blakeney Point.

233

Charles Rothschild.

reserve, even by modern standards, but only contained the distal end of the point. Securing the remainder of the shingle spit and marshes would have to wait.

Rothschild had both the wealth and the motivation to make generous gestures. He was senior partner to the bank of N M Rothschild & Sons as well as being a skilled and energetic naturalist. Rothschild moved amongst the elite, with leading statesmen, academics and businessmen in his acquaintance. Over the next few years he would play a pivotal role in the development of nature conservation in Britain.

Rothschild founded the Society for the Promotion of Nature Reserves in the year following the purchase of Blakeney. He and his associates believed it was necessary to secure examples of land throughout the country that 'retain primitive conditions and contain rare and local species liable to extinction owing to building, draining, disafforestation, or in consequence of the cupidity of collectors'. Whilst the language of 1912 is somewhat dated, the sentiments are admirable.

There then followed extensive correspondence with naturalists from all over the kingdom to collate a list of sites 'Worthy of Preservation'. The outbreak of war in 1914 gave a sense of urgency to their work as the government was asking county councils to identify waste ground suitable for improvement as part of the war effort. Waste ground capable of agricultural improvement was the very stuff of Rothschild's review. By 1915 he was in a position to present the list of proposed sites to the government's Agricultural Board. He looked to the Board to recognise the nature reserves and to divert attempts to improve agricultural productivity away from them. He also hoped that through the Board he could engage with the military who had a significant interest in unproductive land for training. The response of the Board was courteously ambiguous; there were mild assurances but little else. Nature conservation was not one of the roles of the Board in particular, nor government in general.

Rothschild's list of 284 sites 'Worthy of Preservation' included 29 selected specifically for their saltmarshes. Saltmarshes, or in Rothschild's terms 'Saltings', had been highlighted early in the process as of especial interest, as were their associated sand dunes and shingle beaches. The distribution of sites around the country reflected the enthusiasms and knowledge of the correspondents. No saltmarshes were selected in Wales or Ireland but the habitat was coincidentally included in sites selected for other interests such as the dunes systems

of Kenfig and Aberffraw in Wales together with Rostonstown Burrow and Raven's Point, now both in Éire.

Only three saltmarsh sites were selected from Scotland, all of which are to be found in the north-eastern counties of Ross-shire, Caithness and Sutherland. The Shin River and the Tain Sandhills are respectively found on the tidal headwaters and mouth of the Dornoch Firth. Further north the Wick River drains the blanket bogs of the Flow Country to reach the sea by way of brackish marshes.

The Thames estuary together with the coasts of East Anglia and Lincolnshire account for the majority of the English sites. The Solent is also well represented but otherwise there are just two West Country marshes, Porlock Weir and the estuarine elements of Braunton Burrows, together with outlying sites at Lindisfarne and Romney Marsh.

With the government's refusal to engage with safeguarding nature reserves it fell to benevolent individuals and charitable societies to progress Rothschild's vision. The work of the Society for the Promotion of Nature Reserves lost momentum with Rothschild's declining health and death in 1923. However, the process of

Calshot Bank and Saltings, one of Rothschild's saltmarshes 'Worthy of Preservation', have been heavily developed since his recommendation.

Tain Sandhills are substantially intact and are used by the RAF for live firing.

purchasing nature reserves which began at Blakeney made a faltering start before the onset of the Great Depression and the gathering clouds of another European war.

Three saltmarsh nature reserves

The initial purchase of Blakeney Point was intended to be the first step in securing a significant stretch of the North Norfolk Coast as a nature reserve. The original reserve was expanded westwards in 1923, with the National Trust acquiring the saltmarshes and dunes of Scolt Head; however, when the marshes at Cley and Salthouse to the east came on the market they declined to take an interest. Francis Oliver was an enthusiastic member of both the Society for the Promotion of Nature Reserves and the National Trust. The unwillingness of the National Trust to acquire Cley and Salthouse, and the inability of the Society to do so, led him to support the founders of the Norfolk Naturalists' Trust, who secured the purchase in 1926.

Charles Rothschild's saltmarshes 'Worthy of Preservation' c.1915

Site name	Location
England	
Holy Island (Northumberland)	**North-east England**
Orford (Suffolk)	**East Anglia and eastern England**
Hemley: River Deben (Suffolk)	
Blakeney Point (Norfolk)	
Burnham Overy (Norfolk)	
Broads (Norfolk)	
Skegness (Lincolnshire)	
Thames Estuary (Essex/Kent)	**Wider Thames estuary**
Shoeburyness (Essex)	
Benfleet (Essex)	
Canvey Island (Essex)	
Hall Marshes (Essex)	
Osea Island (Essex)	
St Osyths (Essex)	
Ray Island (Essex)	
The Medway (Kent)	
Whitstable (Kent)	
Pegwell Bay (Kent)	**South-east England**
Sandwich (Kent)	
Woolpack and other Fleets: Romney Marsh (Kent)	
Pagham Harbour (Sussex)	**The Solent**
Chichester Harbour (Sussex, Hampshire)	
Langstone Harbour (Hampshire)	
Calshot Bank and Saltings (Hampshire)	
Porlock Weir (Somerset)	**English West Country**
Braunton (Devon)	
Scotland	
Tain Sandhills (Ross-shire)	
Shin River (Sutherland)	
Wick River (Caithness)	

Source: adapted from The Rothschild Reserves Archive (www.wildlifetrusts.org/Rothschild-Reserves)

The Norfolk Naturalists' Trust was the first of the independent local trusts to emerge from Rothschild's initiatives who now coordinate their work through the partnership of Wildlife Trusts.

With hindsight, Francis Olivier was studying the Norfolk coast at a particularly interesting stage in its ecological development. Whilst Oliver viewed the marshes as being in a natural primitive condition, their character reflected local history and traditions. When Oliver first knew them the marshes had until relatively recently been a component of an agricultural system based on big estates and common rights. Postgate (1973) and Williamson (2006) describe pre-enclosure marsh-edge villages situated by freshwater springs above the reach of the tides. The higher ground above the villages was capped with heathland that had been used for grazing and gathering fuel as well as periodically cultivated brecks. Until enclosure the arable fields around the villages were managed in common as open field systems with communal grazing rights over the shackage once the harvest had been taken. The coastal marshes offered extensive areas of common grazing as well as the ability to gather wild food by fishing, wildfowling and harvesting samphire *Salicornia* spp.

By the early 20th century the provisions of the Enclosure Acts of the previous century had been implemented. The agricultural economy had shifted from being based on commons to one founded on enclosed farmland. The intertidal marshes were sometimes the exception to enclosure and the associated extinguishment of common rights. In cases such as at Brancaster, Holme-next-the-Sea and Burnham Overy the marshes remained legally unenclosed and the traditional common rights were retained.

The everyday history of the commonplace is seldom a matter of official record. How common rights were exercised along the Norfolk coast during the early 20th century is not reported in Oliver's or any other ecological studies. An exception was Arthur Tansley who made a brief comment regarding sheep at Scolt Head and Blakeney. In what appears an almost throwaway comment in 1939 he recognised that the plant communities were determined largely by the grazing of sheep, but then offered no further comment or analysis. The photographs accompanying accounts contemporary to Oliver's studies do not illustrate livestock but the vegetation is remarkably short compared with today. Descriptions of the vegetation to the south of the marsh at Blakeney from the 1920s and 1930s are also consistent with grazed open conditions, unlike the Sea-purslane *Atriplex portulacoides*

Cley Marshes.

communities that now dominate much of the marsh. Conversations I have had with local residents suggest that the grazing of cattle and horses on the marshes around Blakeney died out 'sometime between the wars'. Other accounts report that sheep grazing persisted into the mid-20th century. Petch and Swann's *West Norfolk Plants Today* (1962) stated that mature saltmarshes dominated by Common Saltmarsh-grass *Puccinellia maritima* provided grazing for large numbers of sheep. Even so, sheep were not the only livestock along this coast during this period. In their study of commons and village greens Denman *et al.* (1967) describe cattle grazing the 280ha common between Thornham and Titchwell, sheep and horses being prohibited by ancient custom. The marsh carried no more than forty beasts, about two-thirds of its extent being intertidal mud with no forage value.

Traditions of grazing the common marshes of Norfolk survived up to the turn of this century. I recall staying at Brancaster between Blakeney and Scolt Head in the late 1990s where the marshes are part of an extensive common. The commoners' ponies pressed up against the windows of our cottage as high tides lapped around their fetlocks. I walked that marsh again in 2015; the ponies had gone, and all signs of the working common were gone. What had been a brackish marsh of subtle diversity was reduced to an upsettingly impenetrable mass of Common Reed *Phragmites australis* and Common Nettle *Urtica dioica*.

There were two other purchases of saltmarsh nature reserves in the 1920s, both with similarly chequered histories.

Rothschild's enthusiasm for buying land was not diminished by his illness in later life. When 40ha of marshes at Ray Island on the Blackwater Estuary in Essex came on the market he offered his Society £1,000 towards their purchase. Ray Island was one of the sites previously identified as worthy of preservation as a reputed location of the Essex Emerald moth *Thetidia smaragdaria* and as a place that was completely isolated by creeks from other properties. The governing committee of the Society declined his offer of funds. A botanical investigation had advised them that, despite it being on their list, the land was of insufficient interest. Having been rebuffed, Rothschild proceeded with the purchase in a private capacity, securing the property in 1920 after protracted negotiations. After his death in 1923 the land was offered to the National Trust who declined to accept it as they judged it lacked landscape beauty or historical interest. Despite their earlier rejection the land was transfered to Rothschild's Society who, following advice from Rothschild's widow, Rozsika, sold the island and its marshes in 1925 in order to have the funds to better manage his other bequest of Woodwalton Fen.

Perceptions of value change, and when Ray Island came back on the market in 1970 it was bought by the National Trust and is currently managed on their behalf by the Essex Wildlife Trust, the local successors of Rothschild's Society. In the meantime, the farm on the island had run derelict and the marsh pastures where Essex Emeralds once fed on Sea Wormwood *Artemisia maritima* had tumbled down into scrub. It now falls to this generation of nature conservationists to recover the exceptional qualities that Rothschild recognised in the place.

The third and smallest saltmarsh nature reserve founded in this period was also in southern England, in this case in the heart of the Romney Marshes. The Royal Society for the Protection of Birds was at first a reluctant landowner. The priorities of the Society during this period reflected their development from campaigners against the trade in exotic fur and feathers. In 1928 they acquired their first property, a

The Essex Emerald moth is extinct in Britain.

field at Cheyne Court, part of Walland's Marsh. The Cheyne Court reserve was small at some 7ha. The habitat was a relic of the extensive brackish grasslands that typified Romney Marsh prior to agricultural intensification. These marshes had potential for ground-nesting birds as well as for wildfowl that grazed on the winter-flooded pastures. A curiously small fragment of Walland's Marsh was included in Rothschild's list as the 'Woolpack and other Fleets'; Cheyne Court was a part of the same marshes but outside Rothschild's boundary.

The fate of the RSPB's first nature reserve was bound up with the surrounding landscape. Pumps were installed to drain the core of Romney Marsh in the 1940s, with the last floods covering the land in 1960. In 1950, Cheyne Court was sold as it was too small and too prejudiced to function as a nature reserve. The habitat retained some value sufficient for it to be notified as a Site of Special Scientific Interest, if not for birds then for the aquatic wildlife of the dykes. This statutory designation proved ineffective and in 1982 60ha of Walland's Marsh was ploughed.

In the late 1930s the vast majority of the saltmarshes and the other habitats that Rothschild aspired to secure as nature reserves remained unprotected. The more rural sites survived unmodified as traditional agricultural practices contributed to their maintenance. In the pressured estuaries of southern England the agricultural depression reinforced a perception that saltmarshes were worthless; the landscapes worthy of preservation proved convenient dumping grounds for domestic and industrial waste. At the outbreak of war the future of Rothschild's vision was far from promising.

Modest proposals chapter fifteen

T o some, the 1940s are not history but the stuff of memory. In the span of a single lifetime nature conservation has become a dominant land use over swathes of Britain's coast, including its saltmarshes. This is a tale of two sectors, of the State and of volunteers.

Archived papers of the Royal Society of Wildlife Trusts contain an eclectic range of documents dating from their origins as the Society for the Promotion of Nature Reserves. Amongst these papers are memoranda from a standing Conference on Nature Preservation in Post-war Reconstruction. Between 1941 and 1945 the Society gathered together leading figures from learned societies and local government to consider the fate of nature; when all could be lost they were inspired to consider what they were fighting for.

The inaugural conference was held in spring 1941 as the British Army was evacuated from Crete. Later that year the Society's first memorandum was issued as Britain stood alone with Europe fallen, north Africa falling and Russia under siege. Despite everything there was still hope enough to plan for a better future. By 1942 the United States of America had entered the war and the Paymaster General was appointed by the British cabinet to co-ordinate plans for reconstruction. He met with and approved the work of the conference, seeking their advice on the place of nature reserves in any general scheme of national planning. Through the initiative of the standing conference nature conservation was quietly absorbed into the emerging vision for a modern Britain. The spirit of the age is reflected in the decision of the Ministry of Town and Country

OPPOSITE PAGE:
Loch Gruinart, Islay, an
RSPB nature reserve.

Planning White Paper (1947) to quote verbatim from the conclusion of the Nature Reserves Investigation Committee:

After the long years of death and destruction, these modest proposals are unhesitatingly commended as a first step to the renewed study and appreciation of life.

Establishing the conservation estate

The statutory sector

The vision for a reconstructed Britain was of a modern age defined by rational thought and where there was betterment for everyone. Wildlife and a beautiful countryside were to have a place in this new world. The wartime conservation debates concluded between 1947 and 1949. Through parallel processes a special committee for Scotland and another for England and Wales set out their recommendations for nature conservation in a series of White Papers.

The strategy to conserve Britain's wildlife had three elements, all of which related to how land was to be managed. Those elements were National Nature Reserves and Conservation Areas together with Sites of Special Scientific Interest.

National Nature Reserves were proposed as places owned or otherwise managed as sanctuaries by the State which would be kept under constant and close scientific scrutiny. Between them the Committees proposed 97 sites; of these, 24 contained areas of saltmarsh including perched saltmarsh communities, brackish lagoons and grazing marshes.

In addition to these discrete research-based nature reserves there were proposals to establish bigger Conservation Areas, alternatively called Scientific Areas in England and Wales. These were places to be conserved on a landscape scale, the greatest of which was the entire north-west coast of Scotland, so big it was given a special category of its own. In these areas it was intended that the State exercise a regulatory influence on land use change. In Scotland only one Conservation Area was promoted that was substantially saltmarsh; this was on South Uist, which was selected on the bizarre criterion that its chief characteristic was a paucity of southern species.

The third element of the strategy was to identify smaller sites of considerable biological or geological importance and to advise the planning authorities and their owners of their existence.

National Nature Reserves and Conservation Areas supporting saltmarshes as recommended to government 1947–49

Scotland			
Name	**Location**	**Area**	**Status in 2016**
National Nature Reserves			
NNR 1 Hermaness	Shetland	858ha	NNR established
NNR 2 Noss by Bressay	Shetland	300ha	NNR established
NNR 3 Bay of Nigg	Cromarty Firth	77ha	RSPB nature reserve
NNR 7 Estuary of the Ythan	Aberdeenshire	1,177ha	NNR established
NNR 10* Gruinart Loch and Flats	Islay, Inner Hebrides	788ha	RSPB nature reserve
NNR 11 St Cyrus	Kincardineshire	387ha	NNR established
NNR 12 Reach of Tay estuary including Mugdrum Island	Perthshire and Fife	169ha	Tay Valley Wildfowlers Association Nature Reserve
NNR 13 Part of Tentsmuir including Morton and Morton Lochs	Fife	368ha	NNR established
NNR 16 Coastal strip Aberlady–Gullane	East Lothian	207ha	LNR established
NNR 19* Reach of Solway between rivers Lochar and Nith	Dumfriesshire	1,742ha	NNR established Wildfowl and Wetlands Trust nature reserve
NNR 22 Newton Estate	North Uist, Outer Hebrides	5,662ha	
NNR 23 Loch Druidibeg	South Uist, Outer Hebrides	828ha	NNR established
Conservation Areas			
NCA 22* Saltmarsh at Balgarva	South Uist, Outer Hebrides	65ha	–
Special Conservation Area North-west Sutherland	–	204,982ha	–

England and Wales			
National Nature Reserves			
NR.4 Roudsea Wood	North Lancashire	260ha	NNR established
NR.14 Newborough Dunes and Llanddwyn Island	Anglesey	400ha	NNR established
NR.21 Kenfig Dunes	Glamorgan	600ha	NNR and LNR established
NR.28 Heaths from Studland to Arne	Dorset	1,200ha	Part NNR, part RSPB, part National Trust

National Nature Reserves cont.			
Name	**Location**	**Area**	**Status in 2016**
NR.29*	Hurst Castle and Keyhaven	Hampshire	365ha	Wildlife Trust nature reserve
NR.35*	High Halstow Marshes	Kent	770ha	Minority part NNR. Part RSPB nature reserve
NR.37	Deal Sand-hills	Kent	100ha	NNR established. Wildlife Trust nature reserve
NR.52*	Horsey Island	Essex	770ha	NNR established. Part Wildlife Trust nature reserve
NR.54	Minsmere Level	Suffolk	890ha	RSPB nature reserve
NR.58	Hickling Broad and Horsey Mere	Norfolk	1,035ha	Wildlife Trust and National Trust NNR
NR.60*	Blakeney Point	Norfolk	445ha	NNR established National Trust
NR.61*	Scolt Head Island	Norfolk	325ha	NNR established National Trust
Scientific Areas			
SA.3	Lake District	Cumberland, Westmorland and Lancashire	231,000ha	
SA.8*	Artro Valley and the Rhinogs	Merionethshire	4,140ha	
SA.9	Pembrokeshire Coastline		10,350ha	
SA.10*	The Gower	Glamorgan	12,430ha	
SA.13	Lizard Peninsula	Cornwall	15,500ha	
SA.15	Berry Head to Start Point	Devonshire	2,600ha	
SA.16	Sidmouth to White Nothe	Devonshire and Dorset	6,500ha	
SA.17*	Isle of Purbeck	Dorset	18,100ha	
SA.19*	New Forest	Hampshire and Wiltshire	63,000ha	
SA.25*	Dungeness	Sussex and Kent	13,000ha	
SA.29*	Suffolk Coast and Heaths		33,000ha	
SA.31	The Broads	Suffolk and Norfolk	47,000ha	
SA.32*	North Norfolk Coast		11,400ha	

In most of the Conservation Areas/Scientific Areas the presence of saltmarsh is coincidental to the primary reason for selection. An asterisk * identifies those sites where saltmarsh is a significant element.
NNR: National Nature Reserve
LNR: Local Nature Reserve   Sources: Ritchie (1949) & Ministry of Town and Country Planning (1947).

It was envisaged that there would be many hundreds of such sites and, having been informed, the planners and landowners would voluntarily make provision to safeguard them; these sites were to be known as Sites of Special Scientific Interest.

Visionary White Papers are diluted by the reality of legislation, budgets and politics. The concept of Conservation Areas fell on stony ground, with dire consequences for British wildlife. From the early 1950s a rolling programme of surveys identified Sites of Special Scientific Interest that were duly notified to planning authorities. Liaison with landowners proved more difficult as in many cases they were unknown to the officers making the notifications. National Nature Reserves (NNRs) were established under The National Parks and Access to the Countryside Act 1949 with the statutory purpose for 'preserving flora, fauna or geological or physiographical features of special interest in the area and/or for providing opportunities for the study of, and research into, those features'.

Over the years since the 1949 Act the original concept of National Nature Reserves has been diluted through an expanding suite of strategic objectives. The broader objectives and unforeseen scale of destruction of Britain's wildlife led to many more sites being promoted as worthy of NNR status than was originally planned. The post-war schedules were reviewed in 1955 and again in 1965 with the results of the latter being edited by Derek Ratcliffe and published in 1977 as *A Nature Conservation Review*. The review started out as an analysis of the diversity of habitats in Britain with the intention of updating the lists of proposed nature reserves for acquisition by the State. During its production the review morphed into something rather different. By the time it was published it represented an aspirational list of exceptional habitats to be conserved by whatever means may prove most appropriate. The rationale behind site selection had evolved from the previous scientific and educational approach to comprise ten 'Ratcliffe criteria' of size (extent), diversity, naturalness, rarity, fragility, typicalness, recorded history, position in an ecological/geographical unit and potential value, together with what the scientifically minded Ratcliffe described as the awkward philosophical matter of intrinsic appeal. In explaining his rationale he emphasised that even in 1949 the selection of sites was not purely academic but also sought to serve 'that considerable section of the public who without any scientific interests can derive great pleasure from the peaceful contemplation of nature'.

Nature Conservation Review sites supporting saltmarshes and brackish wetlands

	NCR Grade	
Wales		
Oxwich	I	**The Gower**
Burry Inlet	I	
Newborough Warren-Ynys Llanddwyn	I	**Anglesey**
Cors Fochno (Borth Bog)	I*	**Cardigan Bay**
Dyfi	I	
Glannau Harlech	I	
Tywyn Gwendraeth: Towyn Burrows	2	
England		
Dee Estuary	I*	**The Wirral and Welsh borders**
Ribble Estuary	I*	**North-west England**
Morecambe Bay	I*	
Walney and Sandscale Dunes	I	
Duddon Sands	2	
Drigg Point	2	
Lindisfarne	I	**North-east England**
Teesmouth Flats and Marshes	2	
North Norfolk Coast	I*	**East Anglia and eastern England**
Hickling Broad	I*	
Humber Flats and Marshes	I*	
The Wash Flats and Marshes	I*	
Stour Estuary	I	
Orfordness/Halvergate	I	
Upton Broad	I	
Saltfleetby/Theddlethorpe	I	

Blackwater Flats and Marshes	I*	
Foulness and Maplin Sands	I*	
Hamford Water	I	
Burntwick Island/Chetney Marshes (Medway Marshes)	I	**Wider Thames estuary**
The Swale	I	
Leigh Marsh	2	
High Halstow/Cliffe Marshes	2	
Allhallows Marshes/Yantlet Creek	2	

Dungeness	I*	
Cuckmere Haven	I	**South-east England**
Sandwich/Pegwell Bay	I	

Chichester/Langstone Harbours	I*	
North Solent Marshes	I	**The Solent**
Newtown Harbour	2	

Bridgwater Bay	I*	**Severn estuary**
Berrow Marsh	I	

Poole Harbour	I	
Exe Estuary	2	**English West Country**
Fal-Ruan Estuary	2	

Scotland		
Upper Solway Flats and Marshes	I*	**South-west Scotland**
Wigtown Bay	2	

Loch Gruinart-Loch Indaal	I*	
Loch An Duin	I*	
Grogarry Lochs	I*	**Hebrides**
South Uist Machair	I	
Northton, Harris	2	
Loch Nam Feithean	2	

	NCR Grade	
Scotland cont.		
Noss	I*	
Hermaness	I*	
Mousa	I	**Shetland and Orkney**
Lochs of Harry and Stenness	I	
Haaf Gruney	2	
Lochs of Spiggie and Brow	2	
Wick River Marshes	I	
Culbin Sands	I	
Invernaver	I*	**Highland coast**
Loch Fleet	I	
Morrich More	I*	
Ruel Estuary	2	
Tay Estuary	I*	
Sands of Forvie	I*	
Loch of Strathbeg	I	**Scottish east coast**
St Cyrus	I	
Montrose Basin	I	
Eden Estuary	2	

Grade 1. Sites of international or national importance. The addition of * indicates a site of international importance.
Grade 2. Sites of equivalent or only slightly inferior merit to Grade 1. Potential alternative sites should it be impossible to safeguard Grade 1 sites.
Grade 3. Sites of high regional importance. Some sites in this grade have a lower rating if located in a region with more extensive natural ecosystems.

Source: Ratcliffe 1977. Adapted from *A Nature Conservation Review*, edited by Derek Ratcliffe, 1977, © NERC © NCC 1977, used with permission from the Joint Nature Conservation Committee.

By the late 1970s it was clear that nature reserves alone could not stem the loss of that great pleasure from Britain's countryside.

The vision for National Nature Reserves proved difficult to deliver. In many cases the tenure over the land negotiated by the relevant government agency, the Nature Conservancy (1949–73) and its successor the Nature Conservancy Council (1973–1991), fell short of being able to determine how the land was to be managed. The ambition to maintain nature reserves under close and constant scientific scrutiny similarly proved elusive. By the early 1980s the statutory sector was losing momentum in establishing and delivering NNRs. On paper, the numbers of NNRs continued to rise as from 1981 any land managed by an approved body could share that prestigious title. Despite numerous difficulties, the saltmarsh NNRs proposed in the 1940s have mostly, at least in part, been brought under conservation management. The exceptions represent both ends of the conservation spectrum. Newton Estate in the Outer Hebrides remains under the sympathetic management of machair crofting and it is questionable whether NNR status would bring any additional benefits. In contrast,

Halstow Marshes, the speculative location for a new London airport.

the Halstow Marshes of Kent are subject to development pressures including aspirations for a third London airport.

Today the network of National Nature Reserves in Britain includes some 50 sites where saltmarshes are present. These range from the perched saltmarsh communities of Shetland's cliffs to the extensive intertidal flats of the Wash. The inclusion of perched saltmarshes is due to the fortunate coincidence of the habitat with bird-breeding cliffs. In contrast, saltmarsh NNRs on lowland coasts have a tendency to reflect intertidal habitats having little connectivity with terrestrial transitions. The NNR series includes some brackish aquatic habitats but completely excludes inland saltmarshes, with hindsight an understandable omission given that their importance was not appreciated until relatively recently.

The majority of National Nature Reserves, by number, are in England; the numerical dominance is as much the result of the highly fragmented habitats of the English coast as it is of scale. By both number and area the majority of England's saltmarsh NNRs are concentrated along the south-east coast from the wider Thames estuary to the Humber with significant outliers around the Solent, Lindisfarne, the Ribble estuary and Bridgwater Bay. The Welsh series of NNRs tend to support saltmarshes where they are a minority part of much larger sand-dune complexes such as at Merthyr Mawr and Newborough Warren. There are extensive marshes and intertidal flats in the NNRs at Morfa Harlech and Whiteford, together with the estuary of the Dyfi. In Scotland saltmarshes are similarly coincidental with more extensive coastal habitats. The largest substantial saltmarsh NNRs are at Caerlaverock and Loch Fleet, with intertidal habitats providing coastal transitions to bog at Mòine Mòhr, and ancient woodland at Taynish.

The work of central government and its agencies has been complemented by local government. Circumstances across Britain vary but there are many coastlines where important areas of saltmarsh have been secured through ownership by local authorities. Councillors empowered by the 1949 Act could declare land they controlled to be Local Nature Reserves. The Act defines such nature reserves as land that is managed solely for a conservation purpose, or, if the management of the land includes a recreational purpose, that use does not compromise its management for the conservation purpose.

Local Nature Reserves (LNRs) by their very nature reflect local priorities; as such there is an uneven distribution of LNRs.

Britain's largest LNR covers nearly 3,000ha of the saltmarshes and mudflats of Wigtown Bay in the outer Solway. The designation was a useful mechanism to establish a local partnership to oversee the management of the marshes. Such extensive designations are rare but can also be found in Foryd Bay near Caernarfon and on the Kingsbridge estuary in Devon.

Britain's largest Local Nature Reserve covers the estuary of the Cree at Wigtown, Dumfries and Galloway.

There is a notable concentration of saltmarsh LNRs in the estuaries of central-southern England between Pagham Harbour in Sussex through the Solent to Christchurch Harbour in the west. Within this short stretch of coast there are 24 LNRs, more than all the other English saltmarsh LNRs put together. This concentration reflects a period in the late 20th century when local and national government worked together to establish a comprehensive scheme of conservation management of the Solent coast.

In practice, many LNRs are public open spaces that through designation have been secured against development. By such means local politicians secure their legacy whilst restricting the opportunities of their successors. When they were first conceived, 'local' meant just that, places of local importance. In reality the majority of saltmarsh LNRs are also designated as of national or even international importance.

The voluntary sector

Having effectively lobbied government to establish nature reserves, the voluntary sector did not withdraw from acquiring land itself. The history of acquisition by the voluntary sector is one of opportunism and strategy tempered by pragmatism. Whilst being mindful of the agenda of the statutory sector, the voluntary conservation bodies have been free to respond to their own priorities.

The principal voluntary bodies engaged with the conservation of Britain's saltmarshes were all established before the passing of the 1949 Act. To some, such as the National Trusts, nature conservation is a part of broader ambitions to safeguard the nations' heritage. The Royal Society for the Protection of Birds (RSPB) has a particular interest in coastal wetlands due to their importance for birdlife, as does the Wildfowl and Wetlands Trust. Having emerged from Rothschild's Society for the Promotion of Nature Reserves, the partnership of local Wildlife Trusts adopted a broad remit of interest in all of nature. By 1949 county-based Wildlife Trusts were already established in Norfolk (1926), Yorkshire (1946) and Lincolnshire (1948). Today, the Wildlife Trust movement comprises a partnership of 47 locally based charities covering all of Britain as well as Ulster together with the islands of Man and Alderney.

In 1949 Sir Peter Scott's Wildfowl and Wetlands Trust was the most recently established of the conservation charities. Sir Peter set out to develop places where the public could see waterfowl from around the world as well as experience spectacular concentrations of British wildlife. He was an effective mass communicator, using the latest technology of television; he also enjoyed a position in society where he moved amongst the landed gentry. The early nature reserves established by his Trust were therefore partnerships with large country estates, the Trust investing in visitor facilities and site management rather than seeking extensive freehold ownership. Important saltmarshes under the care of the Wildfowl and Wetlands Trust today include Caerlaverock, Llanelli and Washington. The Trust manages two sites on the Severn estuary: the newly rejuvenated saltmarshes at Steart, together with their headquarters at Slimbridge.

The RSPB holds an extensive and highly diverse estate including 4,344ha of saltmarsh spread over 48 of their coastal nature reserves. The largest marshes under their management are on the estuaries of the Dee, Morecambe Bay and the Solway. The RSPB is a major landowner in its own right but also operates as partners and tenants of landed

OPPOSITE PAGE TOP: Astonfields, Staffordshire, is the only inland saltmarsh Local Nature Reserve.

OPPOSITE PAGE BOTTOM: Stanpit Marsh Local Nature Reserve is grazed by New Forest Ponies, Dorset.

Arne, an RSPB nature reserve where saltmarshes grade into heathland, Dorset.

estates and government bodies. For every high-profile visitor attraction such as Titchwell or the Newport Wetlands there is a suite of wilder places. As a national organisation the RSPB is well placed to formulate national strategies and direct its resources to priority initiatives.

Numerically, the largest providers of saltmarsh nature reserves in Britain are the Wildlife Trusts, with 90 such reserves. The tenure of Wildlife Trust nature reserves is complex, with many properties held on lease and by agreement as well as through freeholds. As a partnership of local charities the Wildlife Trusts are deeply rooted in the communities they serve; their priorities reflect local needs that are manifest in their saltmarsh nature reserves. As well as caring for extensive estuarine systems the Trusts' portfolios include fragmentary habitats of lesser interest to larger organisations. Their nature reserves therefore range from whole tidal basins such as Montrose and Loch Fleet to the tiny landlocked medieval saltmarsh of Surfleet Lows (Lincolnshire), the 'continental' saltmarsh of Pasturefields (Staffordshire) and the brackish transitions into floodplain mire of Goodwick Moor in Pembrokeshire.

The National Trusts' conservation work is concentrated on the land they own. In a few cases parts of their estates are designated as nature

reserves but it is more usual for their land to be managed without recourse to such labels. The Highland estates of the National Trust for Scotland contain relatively small but important areas of saltmarsh, particularly at Morvich, Kintail and Balmacara. It is in their marsh below the Five Sisters of Kintail that the Saltmarsh Sedge *Carex salina* was discovered in Britain.

The National Trust for Scotland's estate at Kintail.

It is fitting that this review of the voluntary sector should conclude with the National Trust (for England, Wales and Northern Ireland), as they established the first saltmarsh nature reserve in Britain at Blakeney. The National Trust's estate includes some 1,250km of Britain's coastline, over 900km of which has been purchased since 1965 under their Enterprise Neptune campaign. Enterprise Neptune is a long-standing fundraising initiative with the aspiration to buy and protect the nations' unspoilt coastline. This huge estate is concentrated on the least urbanised parts of the coastline with a strong emphasis on the relatively remote and rocky. There are 22 properties with saltmarshes in England, the most extensive holdings occurring on the East Anglian coast, particularly in North Norfolk. At Newtown, on the Isle of Wight, the last predominantly rural estuary of central-southern England is

Saltmarsh grasses colonising the sandflats of the Llanrhidian marshes.

safeguarded through National Trust ownership. In the West Country saltmarshes form an important, if minor, component of the Trust's country house estates such as at Holnicote, Cotehele and Trelissick.

The first property the National Trust acquired under Enterprise Neptune was on the Gower in South Wales. This estate covers much of the southern foreshore of the Burry Inlet including most of the Llanrhidian marshes to Whiteford Burrows. This is the largest of the saltmarshes owned by the Trust in Wales, with other important properties in Pembrokeshire and the sheltered estuaries of Cardigan Bay.

Biologists recognise resilience in diversity, and that dictum is sound in the establishment of saltmarsh nature reserves. There is inevitable competition between conservation bodies for public profile and resources but there are also many examples of practical partnership working. Wildlife does not recognise boundaries applied by administrators, nor do most people who support charities. A flock of Brent Geese passing over the Blackwater marshes of Essex will be relying on habitats provided through the combined efforts of the National Trust, RSPB and Essex Wildlife Trust. The same birds will also be drawing on the resources of Natural England and the military land to the south as well as testing the tolerance of neighbouring arable farmers. In many of our important estuarine systems nature reserves coincide to safeguard meaningfully large complexes of saltmarshes and their associated habitats.

There is no common standard to define a nature reserve; the title reflects intent rather than security. There are many marshes benefiting from management sympathetic to wildlife that bear no nature reserve designation. Such sites are found in the ownership of private individuals and landed estates as well as under the control of the State for purposes such as military training. Saltmarshes around the Wash, the Thames estuary and Morrich More in the Dornoch Firth all benefit from the restrictive necessities of firing ranges. Coast defence engineers in local government and the statutory agencies play an important role in safeguarding saltmarshes as a healthy marsh reduces risk to life and property. In England the Environment Agency has an increasingly important role as a landowner, enabling the better defence of people through securing natural processes. Whilst their core intentions may not be nature conservation, the net result is safeguarding saltmarsh habitats of outstanding importance.

Britain's conservation estate is substantial even if indefinable and unquantifiable. The quality and extent of saltmarshes within sympathetic tenure are testimony to over a century of dedicated effort. In the post-war vision for reconstructing Britain the establishment of nature reserves was just one part of the three-pronged strategy for the rational management of wildlife. Then, as now, most wildlife does not live in nature reserves.

Military training effectively conserves the saltmarshes at Donna Nook, Lincolnshire.

The rise of regulation

Post-war planners never intended nature reserves to be the sole measure to safeguard Britain's wildlife. Regulation by the State of land in private ownership also had a part to play, be it statutory regulation in Conservation Areas or self-regulation in Sites of Special Scientific Interest. The Wash remains Britain's largest saltmarsh complex though much reduced by piecemeal loss; its history illustrates why regulation was needed and how it eventually prevailed.

Gedney Drove End

Saltmarshes lining the embayment of the Wash have been subject to attrition since at least the Roman era. Different authorities have variously calculated that between 32,000 and 42,000ha of saltmarsh are now farmed behind the Wash's seawalls. The mid-point between these two estimates is broadly equivalent to the size of the counties of Rutland or the Isle of Wight. More saltmarshes have been lost from the Wash than from any other estuary complex in Britain.

The nature of saltmarsh formation in the Wash means that not all of this habitat would have been present at the same time. By embanking saltmarshes it is possible to stimulate new marsh growth into the intertidal zone. This response to engineering, and the sheer scale of the Wash, once supported a commonly held notion that the resource was practically limitless. It seemed that any attrition to the marsh, should that be of concern, would be quickly restored; indeed the accretion of new marshes beyond the seawalls presented opportunities for further expansion.

Throughout recent history the prospect of developing the Wash has attracted speculators. In 1839 the engineer of New London

OPPOSITE PAGE:
Storm over Gedney
Drove End.

Bridge, Sir John Rennie, proposed to embank over 60,000ha of the bay. The land would become a new English county with the patriotic intention that it be named Victoria, after the young Queen. His proposal never gained momentum. A century later, in 1930, a consortium promoted a speedway track and assorted sports stadia with parking for 50,000 cars covering over 4,100ha of the shore. This ambition was given short shrift at a public meeting in Boston as it encroached on well-established landed interests. Later that decade Dr J F Schoenfeld, a Dutch engineer in the tradition of Vermuyden, proposed a polder crossing from Snettisham in Norfolk to Wrangle in Lincolnshire, an idea that fell away with the outbreak of war. The optimism of post-war reconstruction brought with it the fantasy of architect Harry Teggin for a city of 750,000 people set in its own farmland with an airport and docks. By 1968 Sir Owen Williams, an engineer grounded in motorways including the notorious Spaghetti Junction, revisited the proposal with the suggestion of building an international airport. At the same time the state-run Water Resources Board proposed to impound the sea falls of the Ouse and Nene to create freshwater reservoirs. The reservoir proposal advanced to a stage of detailed investigations including the construction of two experimental bunds, one on Terrington Marsh and the other an island 3km from the mouth of the Nene. The offshore island is a substantial low hill covering 4ha and is the only land above high tides in the bay of the Wash. By the late 1970s the reservoir scheme was abandoned as impractical and unnecessary, with its island outpost now colonised by a gullery. Lesser Black-backed *Larus fuscus* and Herring Gulls *Larus argentatus* breed within the concave mound of what local people call the Doughnut.

With the exception of experimental bunds, the grandiose schemes of these speculators came to nothing. In contrast, the piecemeal conversion of saltmarshes to agricultural land progressed as opportunities arose. In every one of the first seven decades of the 20th century an embankment was being built somewhere in the Wash to extend the fertile farmlands of the Fens.

Accounts from the mid-20th century reflect how the saltmarshes of the Wash were seen at that time. In his 1946 descriptions of the coasts of England and Wales, Professor Steers described the normality of converting saltmarshes to arable farming under the misnomer of reclamation:

Reclamation is practicable and even easy once a marsh has attained sufficient height. The main needs are an enclosing wall and the construction of a simple drainage system to get rid of salt from the soil: the two pieces of engineering provide a new and valuable addition to the countryside. Reclamation schemes, although very important and quite common, seldom cause much comment.

These attitudes persisted well into the 1960s. In reporting on the future management of saltmarsh commons in 1967, Donald Denman, the Head of the Department of Land Economy at Cambridge, regarded commoning as a custom that had outlived social change and belonged to the past. His account of saltmarsh commons enthusiastically reported on their potential, once drained, to create sweet pastures of exceptional richness prior to reseeding and ultimately conversion to arable.

Henry Clifford Darby was a contemporary of Denman and Steers at Cambridge. His history of the Fens (1940) reflects the assumption that drainage was not only inevitable but also represented admirable progress. The case for drainage was not just economic, it was morally improving. This outlook is demonstrated by the seawall built by the borstal boys at Freiston shore in Lincolnshire. The wall is topped by a monument with the commemoration:

This bank was begun manually by the staff and boys of North Sea Camp 13 March 1936. In this year of 1974 over 500 acres claimed from the sea are ploughed, another 200 acre enclosure is imminent. And plans include a 700 acre strip seaward.

The inscription concludes: 'Quan Quam Malefactors Juvenes illi Patriae Bene Fecerunt'. It is unrecorded whether the education of the imprisoned boys extended to the appreciation of such finely expressed sentiments.

The most recent major embankments of saltmarsh on the Wash were undertaken in the late 1970s. A privately promoted sea defence was built south of Gibraltar Point along a frontage of some 10km. In an exhibition of loyalty the engineering was dubbed the Jubilee Bank in commemoration of the anniversary of the Queen's ascension to the throne. With hindsight, the Jubilee Bank proved to be the last major encroachment into the saltmarshes of the Wash in the 20th century. The place where the tide turned in favour of wildlife was Gedney Drove End.

The journey to Gedney Drove End from King's Lynn takes you into the heart of the silt fen. The medieval village of Gedney sits far inland on comparatively high ground by the former estuary of the Nene; its saltmarsh commons yielded long ago to the schemes of merchant adventurers. A network of minor roads links the village with the sea, tracing those 17th-century seabanks and their successors; on either side the fields testify to the year when the silt-laden tides stopped feeding their growth. Contrary to popular myth, the fens are not flat but stepped; land levels change with staccato abruptness. Passing the now-landlocked lighthouses, the fields become progressively young; 1720 steps up into 1747, then again to 1806 and 1865 with a tidy rounding-off of an awkward curve in the wall in 1978.

Once embanked and drained, the saltmarshes at Gedney Drove End are transformed into productive arable fields.

In February 1980 a public inquiry was held in Spalding to consider the proposal to erect a seawall around 81ha of saltmarsh at Gedney Drove End. That the matter should require resolution through public inquiry was remarkable given the absence of opposition to recent, much larger, schemes. What the case represents is the coming together of a number of themes, institutions and individuals to present an alternative future for saltmarshes.

Since the 1930s local government in Lincolnshire had sought to protect its coastline from overdevelopment. A technical definition of development determines whether a proposal needs planning permission. In the 1970s there was a debate in the County Council as to whether the engineering of seawalls constituted regulated development or was just another agricultural practice and thus free from such administration. The Gedney Drove End proposal was deemed a matter for the District and subsequently County planning committees who recognised its significance and referred the matter to the Secretary of State for determination through a public inquiry.

The case for the construction of the seawall at Gedney was put to the planning inspector as a simple repetition of the then current perceptions of saltmarsh. Newly created agricultural land would add to the local economy, to the wealth of the nation, and any losses to wild nature would be short-lived as the marsh would re-grow. There was an attempt to diminish the nature conservation case by pointing out that the saltmarsh in question was the neighbour of RAF Holbeach, the tranquillity of the shore being grossly disturbed by the bombing and strafing of the adjacent marshes.

The case against the proposed development was presented by the government agency, the Nature Conservancy Council. Its case was founded on its notification to planning authorities that the whole of the Wash was a Site of Special Scientific Interest. For the largest such site in the country the citation of 1972 is a masterpiece of brevity:

The whole area is of exceptional biological interest. The intertidal mudflats and saltmarshes represent one of Britain's most important winter feeding areas for waders and wildfowl outside of the breeding season. Enormous numbers of migrant birds, of international significance, are dependant [sic] on the rich supply of invertebrate food. The saltmarsh and shingle communities are of considerable botanical interest and the mature saltmarsh is a valuable bird breeding zone. In addition the Wash is also very important as a breeding ground for Common Seals.

Given the depth of knowledge and political support held within the voluntary sector, the Nature Conservancy Council presented their evidence to the public inquiry in partnership with the RSPB and the Lincolnshire Trust for Nature Conservation.

The defence of the marsh presented at the public inquiry was wide-ranging, including both the case for wildlife and challenging the economic assumptions of the applicant. The 81ha under debate was portrayed to the Inspector as a component of a much larger suite of habitats. As a local feature the marshes at Gedney Drove End were relatively mature, having formed over the course of the previous century. These older, higher parts of the Wash marshes were free from summer floods and so supported good numbers of ground-nesting birds. Over 40 pairs of breeding Redshank *Tringa totanus* were found within the marsh affected by the proposed development. This was a significant proportion of the breeding Redshank of the Wash due to the particularly favourable conditions; a small take of marsh was argued to have a disproportionate effect on the birds. The international importance of the wetlands was evidenced by the number and diversity of birdlife resorting to the estuary from breeding grounds as far afield as Canada and western Siberia. Any diminution of the habitats that supported the concentrations of waterbirds and wildfowl was argued to have consequences far beyond our own shores.

One of the lessons learned from the experience of the Jubilee Bank was that whilst saltmarshes may re-grow, the lower intertidal area, with its sandflats and mudbanks, did not. Each incremental embankment therefore steepened the profile of the intertidal, 'squeezing' the surviving habitats into progressively unsustainable transitions. In addition to coastal squeeze, the case put before the inspector included a discussion of the consequences of man-induced climate change on sea-level rise, an issue that would become increasingly significant over the following decades.

The issues of attrition to habitats, environmental economics, coastal squeeze and climate change are familiar to anyone considering coast defences today. In the early 1980s the team presenting the Gedney Drove End case were at the forefront of environmental advocacy.

Stuart Crooks, who gave evidence for the Lincolnshire Trust for Nature Conservation, told me of an incident when the inquiry visited the site:

> As we walked along the seabank an RAF jet made a low pass: most of the waders feeding on the mudflats were unperturbed. Some flocks flew up but soon settled. The inspector noticed this. Looking out at the vast numbers of waders and wildfowl, Colin O'Sullivan of the RSPB told the Inspector that this was Lincolnshire's Serengeti.

The Nature Conservancy Council's annual report of 1981 summarised the outcome of the inquiry with the decision of the Secretary of State, Michael Heseltine, who concluded:

> Despite the very strong case for adding to the nation's stock of high quality agricultural land, a sufficiently exceptional need has not been established in this instance to justify encroachment on an area of national and international importance to conservation, and therefore ecological and scientific considerations should prevail.

Following the Gedney Drove End decision the planning authorities in Lincolnshire and Norfolk became confident in applying policies to curtail future encroachments. There were a number of similar schemes under consideration as the inquiry came to its conclusion. The proposal to extend the Jubilee Bank at Wrangle by another 71ha was withdrawn. At Frampton a plan to embank 243ha concluded with the sale of part of that marsh to the RSPB, the remainder being subject to an agreement with the Nature Conservancy Council. There was a scattering of other expressions of interest for works north of Mablethorpe and on the Norfolk shore, but none was progressed; by the end of the 1980s the era of embanking the Wash had ended.

The consequences of the Gedney Drove End decision spread far beyond the shores of the Wash. Over the following years the principles established in the case were cited nationwide to resist numerous acts of attrition. From the 1980s saltmarsh loss through a variety of industrial and leisure developments was, for the most part, successfully resisted.

What was achieved at Gedney Drove End was secured through a closely argued case without recourse to supportive legislation or national policy. If the proposal had emerged a few years later it would have been subject to the provisions of the Wildlife and Countryside Act 1981. This Act became necessary as after 30 years it was painfully obvious that the voluntary principle underpinning the conservation of Sites of Special Scientific Interest wasn't working. The criticisms of government policy came from many directions. Marion Shoard's 1980 polemic *The Theft of the Countryside* grabbed the headlines, whilst in 1983 *Agriculture, the Triumph and the Shame* was quieter in tone but more devastating, having been written by Richard Body, Conservative MP for the fenland constituency of Holland with Boston. The fray was joined by Lord Melchett, a Norfolk farmer with a suitably aristocratic background who re-phrased Marion Shoard's righteous indignation for a wider audience. Britain's countryside was in crisis, tax payers were subsidising the excessive industrialisation of farming, which was generating unwanted surpluses whilst depriving the nation of its best soils, clean water and wildlife.

The 1981 Wildlife and Countryside Act introduced a regulatory system whereby the owner of a Site of Special Scientific Interest had to consult the Nature Conservancy Council before undertaking anything from a list of proscribed activities. If consent was withheld then compensation was due on the basis of profit foregone. Inevitably such opportunities were abused, with the public suitably appalled at having to pay out vast sums to compensate for the loss of an income that itself was derived from subsidy. In its defence, the new regime had the potential to slow down the rate of destruction of the very best places for wildlife. What was frustrating was the inability to use the same funds to produce something positive.

The saltmarshes of Broadland

Whilst the Gedney Drove End case was being heard, the Nature Conservancy Council was engaged in a protracted negotiation over another embayment of the North Sea, the Broads. The outcome of the debate over the future of Broadland was to establish a principle that would be of benefit to wildlife for decades to come. Before exploring the resolution of that dispute it is worth remembering why such efforts were required.

Straddling the counties of Norfolk and Suffolk, the Broads are a complex of wetlands formed from the combined estuaries of three rivers, the Bure, Waveney and Yare. Unlike the Wash, with its unbroken frontage to the North Sea, the estuaries of Broadland formed within an archipelago of islands, dunes and shingle spits. Today the differences in the geography of these embayments are not only physical but perceptual. When you are in the depths of the Broads there is little to remind you that this is land derived from saltmarsh.

Historical accounts of Broadland's wildlife

The earliest accounts of the prodigious birdlife of the Broads date from the mid-17th century and are the work of the polymath Sir Thomas Browne. His numerous writings were anthologised by Simon Wilkin (1835) and reveal how Sir Thomas was interested in everything from urn burial and ostriches to the correction of vulgar errors. His observations on the birds of Norfolk are as much culinary as ornithological. He recounted attending a feast where the Mayor of Norwich entertained the Duke of Norfolk with a dish of Common Cranes *Grus grus*. Grey Herons *Ardea cinerea* were also on the festival menu, as were Bitterns *Botaurus stellaris*, a rather common bird but regarded by Sir Thomas as better eating. To his palate the daintiest of all was Godwit *Limosa* spp., being far superior to Redshank. The flesh of marshland birds could be accompanied by their eggs but these were not prestigious ingredients. Lapwing *Vanellus vanellus* nested in

such plenty that their clutches were taken from Horsey to Norwich by the cartload. Not everything was eaten; Sir Thomas describes Spoonbills *Platalea leucorodia* nesting high in old heronries at Claxton and Reedham. He also recorded White Storks *Ciconia ciconia* around Yarmouth but does not say whether they were breeding. Regardless, both species were shot, the Spoonbill taken not for their meat but for their 'handsomeness'.

An impression of the scale of these wetlands is enhanced through Sir Thomas's records of White-tailed Eagles *Haliaeetus albicilla* and Avocet *Recurvirostra avosetta*. He knew Ruff *Philomachus pugnax* in good numbers throughout the whole of the Yare valley from Norwich to Yarmouth. Winter winds from the north-east brought in mixed flocks of geese (Brent *Branta bernicula* and Barnacle *Branta leucopsis*), swans (probably Bewick's Swans *Cygnus columbianus*) and duck, including Smew *Mergus albellus*, Eurasian Teal *Anas crecca* and Northern Pintail *Anas acuta*.

Common Cranes are returning to nest in Broadland's marshes.

269

Sir Thomas Browne.

Browne knew the Broads at a time when they were less engineered than today. Compared to some other estuaries, such as the Severn, the modification of Broadlands' marshes started comparatively late. There is documentary evidence of flood defences being built in the 13th century and field evidence, such as the embankment of Hickling Priory, suggesting even earlier medieval works. How widespread or effective these early drainage schemes were is open to question; as late as 1486 the King was encouraging initiatives to begin embanking rivers and digging drains in the area of St Benet's Abbey. In contrast to the alluvial deposits of the estuaries, the peatlands at their headwaters were heavily exploited throughout the medieval period. The scale of peat extraction was such to create the interconnected lakes of the eponymous 'Broads'.

Some 200 years after Browne's death an account of Broadland's birdlife was published by the Paget brothers. James Paget was only 20 when in 1834 he and his older brother Charles produced their sketch of the natural history of the neighbourhood of Yarmouth. James was the son of a merchant whose reduced circumstances meant he could not follow his older brothers to Charterhouse and then on to Cambridge. There was enough money to provide him with a basic education in a local school before being apprenticed to Dr Costerton, the local surgeon-apothecary. The botanical skills necessary to a medical professional contributed to his growing knowledge of all aspects of natural history. Yarmouth, despite its wealth of wildlife, could not hold him, and shortly after publishing his sketches he moved to London to study at St Bart's. As his career progressed the observational skills honed on wild nature proved invaluable in advancing medical science. Having excelled in his profession, he was elevated to become Sir James, Baronet, Surgeon-Extraordinary to Queen Victoria.

Young James got to know the Broads at a time of great change. Technological advances in land drainage had by then supplemented horse-powered scoops with wind pumps. Incentives to invest in drainage were constrained by substantial areas of the marshlands being farmed in common. It was pointless to spend capital on works if everyone else would then reap the benefits. Those constraints were lifted in the early years of the 19th century following enclosure, the communally farmed wetlands being converted into private holdings. The brothers' sketch does not refer to enclosure directly but observed that improved banking of the rivers would eventually change the natural character of the district. In time, the Pagets' predictions would come to pass but the Broads they knew were still open to unpredictable flooding. They observed: 'There is no regular boundary to the distance to which the water of the sea is carried up the river by the flood tide, no line can be drawn between them'.

Throughout the introduction to their sketch there is the strong impression that the marshes around Yarmouth were as much salt as fresh with an irregular nature to their salinity. The intertidal basin of Breydon Water was at the core of the saltmarsh habitats, but they extended across the plains of the Yare and Bure well beyond the tidal limits we know today.

Having set the scene, the sketch comprises a series of annotated species lists. As with the records of Sir Thomas, they reflect on the utility of wildlife with an emphasis on confirming the most desirable rarities with a firearm. The abundance of birdlife was indicated through what was sent to market. In the winter of 1829 one trader, a Mr Isaac Harvey, sold 400 wildfowl of various descriptions together with 500 Common Snipe *Gallinago gallinago* and 150 Golden Plover *Pluvialis apricaria*. During the breeding season Harvey made an average weekly shipment of 600–700 Lapwing eggs to London.

Sir James Paget.

Osprey were sufficiently numerous for one or two to be shot most years on Breydon Water and the White-tailed Eagles known to Sir Thomas Browne were seen on seven occasions between 1811 and 1829. In the summer of 1817 a pair of Storks, presumably attempting to breed, were shot on the nearby Burgh Marshes together with a pair of Black-winged Stilts *Himantopus himantopus*, again presumably breeding, shot at Hickling in 1822.

Birdlife was not the only interest of the brothers, so their lists include fish, reptiles, mammals, insects and plants. Botanical records are particularly valuable as plants associated with specialised habitats are indicative of conditions such as vegetation structure and salinity. Divided Sedge *Carex divisa* is a slender-leaved grassy sedge confined to marshes that are distinctly saline. The sketch records this sedge from Acle Bridge, a long way up-river from the outfall of the estuaries. Similarly populations of Pedunculate Sea-purslane *Atriplex pedunculata* were recorded from inland localities such as Runham; this is a species of saline bare ground in grazing marshes. Sea-milkwort *Glaux maritima*, Sea Wormwood *Artemisia maritima* and English Scurvygrass *Cochlearia anglica* were all considered abundant, suggestive of a diverse upper saltmarsh community. Parsley Water-dropwort *Oenanthe lachenalii* was reported as not uncommon, a status that it still enjoys. The brothers' record of Sea Barley *Hordeum marinum* being very common is intriguing, given the exacting habitat requirements of seasonally bare ground in very short saltmarsh pastures. It is difficult to envisage what sort of marshland management would support this grass in such abundance.

Tall herbs of brackish marshes were also known to the brothers including Marsh Sow-thistle *Sonchus palustris* and Marsh-mallow *Althaea officinalis*. They only reported one locality for Marsh Sow-thistle, near Burgh Castle. Today the sow-thistle is much more common, the engineered banks of tidal rivers providing useful extensions to their natural habitats. In contrast, Marsh-mallow was described by them as common in saltmarsh ditch-banks. James would have had a professional interest in the plant as it had value in the apothecary's shop as a thickening agent for ointments. Today the mallow remains widely distributed across the marshes, a hardy perennial that is persistent even when deprived of the brackish conditions within which it originally became established.

Being educated locally with a place in the apothecary trade, James would have had access to the recollections of an older generation of botanists. This knowledge enabled the sketch to include observations on species lost from the area long before his birth; sea-lavenders *Limonium* spp. had not been found 'for the last 50 years', Dittander *Lepidium latifolium* 'was lost long ago', Sea-purslane *Atriplex portulacoides* 'seems now lost' and Sea Clover *Trifolium squamosum* was 'lost long since'.

Why would sea-lavender and Sea-purslane, both common species of saltmarshes, become locally extinct whilst highly specialised species persisted? Today sea-lavender and Sea-purslane are widely distributed around Breydon Water, Dittander is rare and Sea Clover remains absent. Why are there are no records from the Pagets of the commonplace Thrift *Armeria maritima* or Sea Aster *Aster tripolium*? Is it possible that the brothers simply overlooked some species or an error excluded them from their sketch? Sea Aster is currently found scattered across the estuarine soils of the Broads whilst Thrift remains absent.

In most cases it is possible to reconcile historical accounts of natural history with what we know of British habitats today. The species lists provided by the Pagets suggest that some of the saltmarsh communities they knew are without an obvious modern equivalent. The plant that exemplifies this issue was common enough for the Pagets to report it by a local name, Trumpets, *Cineraria palustris*, known to us as Marsh Fleawort *Tephroseris palustris* ssp. *palustris*.

OPPOSITE PAGE:
Dittander (inset) and Sea-lavender have recolonised Broadland since the early 19th century.

Thrift, a curiously long-standing absentee from Broadland.

Trumpets became extinct in Britain in the closing years of the 19th century. In the Broads it grew in the lower reaches of Bure and Waveney valleys together with populations around the rivers Ant and Thurne.

Marsh Fleawort is not an easy plant to overlook as it grows in stands from shin-height up to a metre tall, each erect stem crowned by a fanfare of golden bloom. The flowers have a daisy-like arrangement with seeds carried by the wind like dandelion clocks. The whole plant, particularly when young, can be densely hairy, some specimens to the point of shagginess. Colonies usually establish on open mud but they will grow as emergents should their seedbeds become flooded.

Having become extinct before habitat descriptions became commonplace, we only know the general area where Marsh Fleawort grew in Broadland but not the habitats in which it grew. The fleawort has a worldwide distribution covering much of the higher latitudes of Eurasia and North America. It is not a rare or threatened species; indeed it can become dominant where conditions suit it.

In coastal wetlands in Canada, at equivalent latitude to Broadland, Scott (1995) and Staniforth *et al.* (1998) describe Marsh Fleawort as being associated with exceptionally high levels of goose grazing

Marsh Fleawort, known in
the Broads as 'Trumpets'.

in wetlands ranging from brackish to hyper-saline. The fleawort is
found in sheltered conditions where the tide is insufficiently energetic
to uproot seedlings. Grazing and dunging by geese create fertile but
highly stressed open habitats. Marsh Fleawort is one of a few plants
avoided by the geese and is found growing with other adventitious
annual species including Toad Rush *Juncus bufonius* and Lesser Sea-
spurrey *Spergularia marina*.

Closer to home on the eastern shores of the North Sea, Marsh
Fleawort was studied by Bakker (1960) who investigated botanical
responses to the creation of the Yssel polders following the enclosure
of the Zuiderzee. The fleawort became abundant as one of the first
and most successful colonists of the newly exposed bare muddy
shore. This was attributed to fleawort seeds being dispersed on the

wind, giving the plant an advantage over species with floating seeds or spread through vegetative growth. As in the Canadian habitats, Marsh Fleawort excelled in colonising nutrient-rich mud occupied by high numbers of wildfowl.

Archaeological records from elsewhere in the Netherlands provide an insight into Marsh Fleawort habitats of a previous interglacial, some 44,000–47,000 years ago. A diverse flora was excavated from an ancient silted riverbed near Orvelte in Drenthe. Remains of Marsh Fleawort were accompanied by Cowbane *Cicuta virosa*, another Broadland specialist, in a complex wetland community of fen and brackish species. The excavation was primarily reported by Botterna & Cappers (2003) for the excellence of the remains of Mammoth *Mammuthus primigenius* and Woolly Rhinoceros *Coelodonta antiquitatis*, the ultimate large herbivores of that geological period.

The concluding historical account from Broadland relates back to the Pagets' observations on the unpredictable nature of salt. The River Thurne rises in the marshes of Hickling to flow, by way of the Bure, into the North Sea some 30km downstream at Yarmouth. Contrary to the norm, the headwaters of the Thurne are brackish, with the saline influence decreasing downstream from Heigham Sound to Thurne Mouth.

Beckett & Bull's *Flora of Norfolk* (1999) tells the story of how, in late July 1883, Arthur Bennett treated his daughter to a botanical jaunt up the river. Bennett was a scientifically minded botanist who had spent the previous winter comparing the flora of the Netherlands with England's eastern counties. He compiled a list of plants that he predicted may grow as yet undetected in Norfolk, one of which was the Holly-leaved Naiad *Najas marina*.

Later in life he recounted that July day to the members of the Norfolk and Norwich Natural History Society:

> *I engaged a lad to … take us to Hickling. We passed through Kendall Dyke into the Sounds and dragged for plants finding many Chara species. Passing through Deep Dyke into Whitesley, we took the channel to Hickling near the keeper's house … my daughter at the bow brought up a lot of aquatics with the drag. She passed it to me and I at once saw we had Najas! Three days later Mr H Grove went over the same ground with the same lad and found it again. The lad said to him 'Be you from Lunnon arter weeds? Ar! Yer too late!'*

Holly-leaved Naiads have a cosmopolitan distribution, being found across the world in temperate and tropical latitudes. Throughout its global range the plant exhibits tolerances to a wide range of conditions, particularly clear water with a high calcium content but low nutrient status. The epithet *marina* reflects its capacity to thrive in dilute maritime conditions such as in the Baltic. Towards the northern limits of its range the Holly-leaved Naiad is progressively associated with mildly brackish waters, as described from Hickling on the Thurne.

Bennett refers to species of *Chara*, stoneworts, being collected from the naiad beds. The Thurne catchment is regarded by Nick Stewart (2004) as the richest place in Britain for stoneworts, with 16 species having been reported from Hickling Broad. The stonewort community there includes Baltic Stonewort *Chara baltica* and Rough Stonewort *Chara aspera*, both associated with mildly brackish conditions. A third saltwater associate, Bearded Stonewort *Chara canescens*, has been known from the system but has yet to reappear following pollution incidents in the 1970s. It is interesting Bennett did not comment on the pondweeds *Potamogeton* spp. growing there, given that he was an acknowledged expert and these waters are particularly rich. His

Hickling Broad.

Holly-leaved Naiad.

satisfaction in finding the boldly toothed fronds of the predicted naiad may have been sufficient for the trip. The material he and his daughter gathered that day is held in the herbarium of Manchester Museum, filed alongside that collected by Henry Groves just a few days later.

The Broads remain the sole location for Holly-leaved Naiad in contemporary Britain. Godwin (1975) describes how its robust bivalve-like fruits are persistent in lake sediments and have been extracted from East Anglian deposits laid down in previous interglacials, the oldest material dating from at least half a million years ago. In more recent millennia there are archaeological remains of Holly-leaved Naiad distributed around the coasts of England and Wales. Its fruits were recovered from a buried marsh by the tidal Ouse in the Cambridgeshire fens where they are reported by Shawcross (1960) as being accompanied by the remains of stoneworts *Chara* spp., Saltmarsh Rush *Juncus gerardii*, glassworts *Salicornia* spp., and Fennel Pondweed *Potamogeton pectinatus*. This brackish marsh community dated from the early Bronze Age, as did the accompanying skeleton of an Auroch *Bos primigenius*.

The origin of the saline waters of Hickling and the Thurne has understandably attracted speculation by naturalists. Early assertions as to the source of the salt quoted by Martin George (1992) included 'the prevalence at times of dense sea fogs which cling to the reedbeds and leave a salt deposit'. In 1907 Robert Gurney speculated that the high salinity of the Thurne 'is due to the existence of salt springs in Horsey Mere, and possibly also to a lesser extent in Hickling Broad'. Whilst rational in intent, no such springs could be found to support the theory.

It fell to Marietta Pallis to offer a plausible explanation. Pallis was a remarkable botanist and ecologist; born in British India of Greek descent, she studied at the University of Liverpool and Newnham College, Cambridge. Her ecological accounts of east Norfolk earned her a place as a member of the Central Committee for the Survey and Study of British Vegetation, the soul female presence on a committee of gentlemen. Pallis made her home at Long Gores Farm at Hickling.

She celebrated her love of swimming and her Greek Orthodox heritage by carving a devotional landscape out of the peat. To mark the anniversary of the fall of Byzantium she excavated a three-quarter-acre swimming pool incorporating the form of a Byzantine eagle with patriarchal crown and crosses.

In 1911 Pallis published an account of salinity in the Broads, noting that the water table at Hickling 'is practically coincident with the level of the sea at low water, and which, moreover, contains most salt close to the sea'. The distance from Hickling to the sea is not insubstantial at some 5km but that land is mostly made up of former estuarine marshes. Subsequent studies have refined Pallis's observations with the identification of layers of variably permeable peats and sands linking the wetlands to the sea.

The entry of salt into the headwaters of the Thurne is currently cryptic and subterranean. There was no such subtlety on the stormy evening of 12 February 1938 when the dunes near Winterton breached and let in the tide. In all, 3,024ha of marsh between Hickling and the sea were flooded, in some places to the depth of nearly 3m; it took three months to plug the gap. The local landowner, Anthony Buxton, was alerted by a neighbour with 'The sea is in, Sir'. Buxton was a respected naturalist and monitored the changes on his estate. Horsey Mere lost its freshwater fish; only Eels *Anguilla anguilla* survived amongst the imported herring, flat-fish and crabs. Within months pioneer species of the saltmarsh were germinating. In an article published that September in *The Spectator* (1938) Buxton reported colonisation by samphire *Salicornia* spp., together with ambiguously named chenopods, sedges and rushes. He described the now barren farmland as reminiscent of the salt deserts of Asia Minor. Common Reeds *Phragmites australis* and Lesser Reed-mace *Typha angustifolia* survived around the Mere, as did Marsh Sow-thistle, which, whilst stunted by the shock, went on to flower. It took five years for salt to be flushed from the system and the land returned to its previous state.

What happened at Hickling in 1938 was a re-opening of a mouth of the Gariensis Ostium, the Great Estuary that defined this part of Norfolk throughout the period of Roman occupation. What is now Yarmouth was then the main entrance to the network of tidal rivers and saltmarshes guarded by the twin forts of Caistor-on-Sea and Burgh. A shallower opening at what are now the headwaters of the Thurne contributed to creating the island of Flegg. The estuary ran

Marietta Pallis dressed in Epirote costume, 1913.

up the Yare valley for some 23km; on a rising tide commercial vessels could trade out of Whitlingham, with smaller boats reaching further upstream to Venta Icenorum, now Caistor St Edmund. Times change; the coastline known to the Romans was relatively young. A thousand years earlier and freshwater was dominant; that too was just a passing phase when subtle changes in sea level combined with drifting dunes and spits to limit the ingress of salt. The estuary outlived the Romans but by the age of Anglo-Saxon kings its mouths were closing again; dunes sealed the channel north of Flegg and a sand spit restricted the entrance at Yarmouth. Whether and when the next changes occur is open to speculation. In the meantime the marshes of the Roman estuary lie behind their seawalls and below the limits of the tide.

The Halvergate debate

It was a recommendation of the planners of post-war Britain that Broadland be declared a National Park to safeguard its exceptional landscape together with being made a Scientific Area in recognition of its importance for wildlife. The government at the time thought otherwise and neither designation was applied; it was a different wartime legacy that came to dominate how public policy and funds were to influence the marshes of the Broads.

In the decades following the war the Ministry of Agriculture, Fisheries and Food administered public investment in improving the agricultural productivity of Britain. An understandable emphasis on food security had grown up from when Britain was besieged during the Battle of the Atlantic. Farmers were recipients of substantial funds which were available not only for capital works but also for guaranteeing commodity prices. Acts of Parliament established institutions to help deliver this investment, including Internal Drainage Boards that were governed by, and could levy rates upon, marshland farmers.

Since enclosure most farming enterprises on the coastal grazing marshes of Broadland have been founded on livestock. There have been periods when arable crops were more profitable and so in the better drained and more accessible marshes there is a tradition of shifting land-use depending on the relative profitability of 'corn and horn'. The availability of substantial grants to the Internal Drainage Boards meant they could invest in upgrading pumps, drains and embankments to reduce the risk of flooding. Some of the larger entrepreneurial farming businesses had established that deep drainage could prepare estuarine soils for arable cropping. Those drains may be well below sea level but the infrastructure provided by the Drainage Boards, supported by public funds, would keep them dry whilst commodity prices guaranteed by the State would keep them profitable.

Through the 1960s and 70s public investment supported a transformation of Broadland's landscape, arable farms expanded and extensive grassland farms were intensified through applications of artificial fertilisers. By the late 1970s this intensification had rendered most of the coastal grazing marshes botanically dull. Saline and brackish habitats persisted around Breydon Water and also between the tidal rivers and their engineered embankments, known locally as Ronds. In the marshes behind those banks only the presence of False Fox-sedge *Carex otrubae* and Meadow Barley *Hordeum secalinum* suggested the possibility of some residual saline influence. A minority of the dykes retained sufficient inflows of saltwater to maintain a brackish flora. Martin George (1992) summarises a study of the marshes of the lower reaches of the Bure and Yare, identifying elevated salt levels in about 20 per cent of the dykes visited, the typical community comprising Fennel Pondweed, Spiked Water-milfoil *Myriophyllum spicatum* and Sea Club-rush. One exceptional dyke by Breydon Water supported tasselweeds *Ruppia* spp., Sea Aster *Aster tripolium* and Annual Sea-blite *Suaeda maritima*.

The birdlife of the marshes proved slightly more resilient to change than the botany. Where intensification had yet to reduce summer water levels there was still an outstandingly rich assemblage of breeding birds. In the late 1970s a survey of the marshes recorded 233 pairs of Redshank, 481 pairs of Lapwing, 68 pairs of Oystercatcher *Haematopus ostralegus*, 15 male drumming Snipe and 457 pairs of Yellow Wagtail *Motacilla flava*. The study area consisted of 10,225ha, of which 1,720ha were under cultivation. The combination of extensive grasslands with some ploughland was particularly productive for birds needing short areas for nesting with adjacent grasslands for feeding.

Similarly, where shallow winter flooding persisted it continued to attract wildfowl including Pink-footed *Anser brachyrhynchus*, White-fronted *Anser albifrons* and Bean Geese *Anser fabalis* along with Bewick's Swans and Eurasian Wigeon *Anas penelope*. The flocks were much smaller than in the recent past and unrecognisable from those described by Thomas Browne and the Pagets. Even in these reduced circumstances the birdlife of the Broads remained of national importance.

The matter came to a head in 1980 when the Internal Drainage Boards for the marshes south of the Bure announced their intention to upgrade the drainage of what would become known as the Halvergate Marshes. The parish of Halvergate lies at the landward end of a great stretch of marshland covering over 3600ha eastwards to Yarmouth. Halvergate was just one of seven parishes within this landscape which contained over 160 farms.

With hindsight it is interesting that the debate on the future of the marshes concentrated on landscape issues rather than wildlife. The Nature Conservancy Council did not oppose the drainage so long as Sites of Special Scientific Interest were safeguarded; by now the landscape-scale vision for nature promoted by the 1947 White Paper was history. The central government agency with responsibility for landscape conservation was the Countryside Commission. The Commission supported the work of the Broads Authority, a recently established voluntary partnership of local authorities that had adopted a remit remarkably similar to a National Park Authority.

The Broads Authority objected to the proposed drainage as they had previously identified the Halvergate Marshes as of core importance to the landscape of the Broads. Ministers decided they could not authorise public funds to be spent on works until there was a consensus amongst the statutory agencies. The ensuing clash of interests attracted much political and media interest. All parties recognised that what happened at Halvergate would set a precedent for the management of the competing interests of agriculture, wildlife and landscape. The positions of the statutory agencies reflected the contrary and ambiguous nature of government policy. The Ministry of Agriculture was committed to increasing productivity and regarded supporting environmental schemes as outside their legal remit. The Internal Drainage Board appeared puzzled, if not somewhat hurt, at challenges to their intentions. They and the farmers who elected them had grown up in a culture of expectation that the public purse would always be open to them.

The early debates overlapped the passage of the Wildlife and Countryside Bill through Parliament. When the Act became law environmental campaigners tested its effectiveness against the reality of Halvergate. The Ministry of Agriculture continued to subsidise ploughing the marshes as the merits of alternative solutions were debated. The Broads Authority had limited funds to enter into management agreements to retain grassland whilst the Nature Conservancy Council were clear that they could not enter this bidding war and would wait and see what the new legislation would bring. The case was summarised by Martin George (1992), a senior officer of the Nature Conservancy Council at the time, as 'complex, contentious and time-consuming'. His detailed history of the negotiations is essential reading for anyone interested in the minutiae of how the politics of countryside conservation really works.

After the best part of three years a consensus was emerging that the marshes were worthy of conservation and that some means should be found to encourage livestock enterprises. It remained unresolved which government body would meet the cost and on what basis payments would be made. The option of the State compensating farmers for not receiving public subsidies whilst requiring nothing in return received a hostile reception by the press and public. The exposure of State subsidies, the extraordinary economics of compensation and imbalanced calculations of profits foregone were all fuel for the public debate. The cost to the public purse of one government body confounding the aspirations of another had resulted in what Lord Buxton (1981) described as 'A recipe for conflict and taxpayers' involuntary profligacy'.

It took until March 1985 before an experimental package was launched that was acceptable to all; this was the Broads Grazing Marsh Conservation Scheme. The sums involved were realistic, entry into the scheme required compliance with simple criteria and the whole process was managed under an admirably lean administration. Within a year most eligible marsh farmers had joined up. By 1987 the experiment was given a more permanent basis and by 1998 the scheme involved over 900 farmers covering 18,000ha of grazing marsh. As with the case of Gedney Drove End, it was local practice that set a national precedent. The Halvergate scheme evolved into the diversion of agricultural subsidies from the single objective of production into delivering a range of social and environmental benefits. The initial motivation behind the scheme was landscape conservation, of importance in its own right and essential in supporting the tourism economy of the Broads. Successor schemes remain the cornerstone of safeguarding wildlife in Britain's farmed habitats.

After 30 years it is possible to review what was achieved by this redirection of public money. By 1986 some 37 per cent of the marshes had been converted to arable. The Grazing Marsh Conservation Scheme helped bring down the rate of conversion to negligible figures. Subsequent repackaging of the scheme under increasingly complex administrations has provided incentives to reverse the changes, with some 900ha of arable being converted back into grassland since 1995. Even conventionally farmed grazing marsh habitats are biddable when it comes to improving conditions for birds. The manipulation of water levels on arable reversions has succeeded in rebuilding winter flocks of waterfowl; Wigeon and Pink-footed Geese are now present in internationally important numbers. Populations of breeding birds, particularly ground-nesting species, have similarly responded to changing farming methods although they remain concentrated in the marshes under bespoke nature reserve management.

There are too few restorations of brackish grazing marshes from arable to be confident in drawing conclusions for species other than birds. Further south on the East Anglian coast at Trimley the Suffolk Wildlife Trust is managing such an arable conversion. The invertebrate colonisation to date is mostly by generalists of wet grasslands. This fauna is likely to diversify as the upper saltmarsh habitat matures. Saltmarsh Rush *Juncus gerardii* swards are becoming

The Share Marshes in Broadland are being restored by the Suffolk Wildlife Trust.

Common Blue Damselfly
Enallagma cyathigerum
resting on Saltmarsh Rush.

Sea Barley ripening
at Trimley.

established on the former arable, with Marsh-mallow *Althaea officinalis* along the ditches and Sea Barley *Hordeum marinum* in broken ground. There is every indication that over time the habitat will continue to diversify.

A combination of regulation and incentives has helped to stabilise the dramatic loss of wildlife through the post-war decades. The very first sentence in a strategic plan adopted by the Broads Authority in 2011 sets out the challenge 'No landscape ever stands still'. Sea levels are rising and what was once saltmarsh will return to saltmarsh in the absence of interventions and investment. Broadland's landscape emerged over the historical period from a Great Estuary. Current coastal alignments are a snapshot in geological time; the processes that drove change in the past have lost none of their vigour.

International
perspectives

For over a century there have been international dimensions to the conservation of saltmarshes and their wildlife. The British government has accepted responsibilities that cross national borders through being a signatory to a series of international treaties and conventions. Local responses to these obligations are helping to secure a future for Britain's saltmarshes.

Treaties and conventions

Theodore Roosevelt was an adventurer and war hero who by dint of the assassination of William McKinley became the 26th and youngest President of the United States of America. When in office, he struck up an unlikely friendship with Sir Edward Grey. Grey was an English gentleman of high birth who believed in the improving qualities of civilisation and despite his dislike for politics was Britain's longest-serving Secretary of State for Foreign Affairs. What these very different men had in common was a love of the natural world together with the conviction that safeguarding wildlife was a role of government.

Much is made of the special relationship between Britain and the United States. The foundation of that relationship has been traced back to Grey and Roosevelt. In 1911, Roosevelt visited England and spent a day walking through Hampshire's countryside with Grey, listening to birdsong. Tradition has it that on their walk they discussed the vulnerability of birds on their migrations across political boundaries. The outcome of their discussion was the 1916 Migratory Birds Treaty, which bound the Dominion of Canada and the United States of America to a common approach to hunting along the coastal wetlands of the Atlantic seaboard.

OPPOSITE PAGE:
The saltmarshes and estuarine woods of the Helford River, Cornwall.

LEFT: Sir Edward Grey.

RIGHT: Theodore Roosevelt.

It would take another generation before international cooperation in nature conservation gained momentum. In the 1970s Britain became signatory to a convention regarding wetlands and migratory species named after the Iranian city of Ramsar. That decade also saw the United Nations convention enabling biological sites to be recognised as part of the World's Heritage. The 1980s introduced conventions developed at Bonn and Bern that reinforced protective measures for migratory species and introduced requirements for the conservation of all forms of wildlife in its natural habitat. The 1990s extended international cooperation into the realm of restoring wildlife habitats on land and sea through the Rio de Janiero Convention on Biological Diversity together with the Convention for Protection of the Marine Environment of the North East Atlantic. The British are signatories to them all.

Site and species protection

The delivery of international obligations is achieved through local provisions. One of the modest proposals of the post-war planners was the establishment of a national network of Sites of Special Scientific Interest (SSSIs). SSSIs remain the building blocks of site protection and are now designated under the Wildlife and Countryside Act 1981, as amended. This Act of Parliament took nature conservation in Britain away from the voluntary principle established in the 1940s to a more regulated regime. Every SSSI has a map defining its precise boundary together with a citation that notifies the owners and statutory bodies as to why that land is of interest. The notification triggers a regime of compulsory consultation whereby if anyone wants to do something on a proscribed list they must first seek the consent of the statutory regulator. That consent may be withheld should the proposed activity be likely to harm the features for which the site is notified; in such circumstances there are measures to recompense the owner by helping fund more appropriate management.

Britain's largest saltmarshes are all notified as SSSIs, with Sutherland & Hill (1995) calculating about 80 per cent of the habitat being covered by this designation. The criteria used for selecting SSSIs have evolved over the decades. In the 1950s the first round of SSSI notification sought to safeguard a representative selection of variation across the nation; the underlying philosophy was similar to selecting National Nature Reserves. From the 1980s the emphasis moved towards resource protection notifying everything above a threshold of size and quality. As originally envisaged in the post-war White Papers, there are hundreds of SSSIs.

Local circumstances have a tendency to determine how the administrative process of SSSI notification gets done. The saltmarshes of the Wash are included in just three SSSIs embracing coastal habitat complexes extending over 63,724ha. In contrast, the smaller and more fragmented saltmarshes of the Solent form a minority part of 9,060ha of intertidal habitats that are spread across 28 different SSSIs. The reality of notification was one of resources; the hugely complex pattern of landownership around the Solent required the process to be completed incrementally.

When it comes to meeting international obligations it is therefore necessary to bundle up parts of numerous SSSIs to enable the regulatory regimes associated with the Ramsar Convention and European Union Directives to be implemented. A map of the administrative layers for conserving Britain's saltmarshes is multi-layered, with diverse boundaries and conservation objectives nested untidily within one another.

The international conservation designation currently relating specifically to saltmarshes arises from the European Union Habitats Directive 92/43/EEC. In the same way as Britain has sought to safeguard a representative selection of habitats across the nation so the Directive seeks to achieve this objective across Europe. The sites selected under this Directive are called Special Areas of Conservation.

There are 60 Special Areas of Conservation in Britain that include saltmarsh habitats (see Appendix B). The features for which these sites are selected reflect pan-European perspectives. Continental saltmarshes are included as 'Inland Salt Meadows', as are the varied saltmarsh types of Europe's Atlantic coast. Perched saltmarshes are present within the designated areas, not by expressed intent but inadvertently as a minor component of the complex of cliff habitats. Prominent in the list are coastal lagoons, the brackish and hyper-saline open waters of saltmarsh ecosystems.

Sea-heath and Sea Aster sward by the Widewater Lagoon.

The conservation of saltmarshes as habitats is complemented by special provisions for protected species. The perilous state of brackish habitats, particularly lagoons, is highlighted by the dominance of their associated organisms on the schedules of protected species. In practice most habitats supporting protected species are already SSSIs; their presence is one of the criteria for selection. The national distribution of protected species is highly localised, with a handful of sites containing most populations. Protected status brings a degree of kudos along with extra administration; whether such status is effective on its own in safeguarding species is open to question. Widewater Lagoon on the Sussex coast is one of a very few sites for a protected lagoon species that is not also an SSSI. This was the only known site in the world for Ivell's Sea-anemone *Edwardsia ivelli*, currently regarded as probably extinct.

Species selected for special protection in Britain tend to be at the edge of their global distribution with much stronger populations elsewhere in the world. Of all these protected species there is only one, the Shore Dock *Rumex rupestris*, that is specifically provided for by the Habitats Directive, a reflection on its global rarity.

Legally protected species found in saltmarshes

Plants		
Pedunculate Sea-purslane	*Atriplex pedunculata*	Saline mud, upper marsh
Slender Centaury	*Centaurium tenuiflorum*	Brackish mudslides, perched saltmarsh
Dwarf Spike-rush	*Eleocharis parvula*	Brackish mudbanks and pools
Least Lettuce	*Lactuca saligna*	Brackish grassland
Welsh Mudwort	*Limosella australis*	Brackish mudbanks and pools
Holly-leaved Naiad	*Najas marina*	Brackish open water
Shore Dock	*Rumex rupestris*	Brackish shingle, perched saltmarsh
Triangular Club-rush	*Schoenoplectus triqueter*	Brackish mudbanks and pools

Algae		
Bearded Stonewort	*Chara canescens*	Brackish lagoons
Foxtail Stonewort	*Lamprothamnium papulosum*	Brackish lagoons

Animals		
Starlet Sea-anemone	*Nematostella vectensis*	Lagoons
Bembridge Beetle	*Paracymus aeneus*	Lagoons
Ivell's Sea-anemone	*Edwardsia ivelli*	Lagoons
A marine hydroid	*Clavopsella navis*	Lagoons
Trembling Sea-mat	*Victorella pavida*	Lagoons
Tentacled Lagoon-worm	*Alkmaria romijni*	Lagoons
Lagoon Sand Shrimp	*Gammarus insensibilis*	Lagoons
De Folin's Lagoon Snail	*Caecum armoricum*	Lagoons and open brackish inlets
Lagoon Sea Slug	*Tenellia adspersa*	Lagoons and open brackish inlets
Lagoon Sandworm	*Armandia cirrhosa*	Lagoons and open brackish inlets
Natterjack Toad	*Epidalea calamita*	Brackish pools
Tadpole Shrimp	*Triops cancriformis*	Brackish pools
Fisher's Estuarine Moth	*Gortyna borelii*	Brackish grassland

This table lists a selection of species protected under the provisions of schedules 5 & 8 of the Wildlife and Countryside Act 1981, as amended by quinquennial reviews. The species included are closely associated with saltmarsh habitats for all of, or critical stages of, their life cycle.

It was birds, notably migratory birds, that established the principles of international cooperation in the conservation of nature. The Ramsar Convention makes provision for a global network of protected wetlands for migratory species which has been supplemented by the European Union Birds Directive 2009/147/EC. There are extensive lists of protected birds, under both British and European provisions, which deal not only with species but also with outstanding assemblages and significant proportions of world populations. No other suite of wild species is so comprehensively addressed.

In theory, the Ramsar Convention seeks to address all wetland wildlife, not just migratory birds. In practice the selection of Ramsar sites is driven by ornithologists, the other interests taking on a supporting role. The provisions agreed at Ramsar have been described to me as a 'scientist's convention', in contrast to the legalistic Birds and Habitats Directives. The citations that accompany designations under the Ramsar Convention use inclusive language which enables protective measures to be applied to species that may have been unknown or under-regarded when the designation was first made. Such an approach is particularly helpful in highly dynamic ecosystems such as saltmarshes. The commitments Britain entered into at Ramsar are met through the application of other measures, with Ramsar sites usually sharing common boundaries with sites identified under the Birds Directive.

The Birds Directive has resulted in a suite of protected sites called Special Protection Areas being established across Britain. The processes for designating Special Protection Areas and Special Areas of Conservation are very similar; indeed, many saltmarsh complexes are subject to both designations. Because of the international importance of Britain's coastal wetlands to migratory birds most of the largest saltmarsh systems are, at least in part, within designated Special Protection Areas. The Birds Directive is not a measure for the conservation of saltmarshes; it coincidentally achieves this in seeking the better management of bird habitats.

During the drafting of this book the referendum into Britain's future as part of the European Union concluded with a majority vote to leave. At the time of writing it is unclear what happens next. If 'leave' means discarding measures derived from the Birds and Habitats Directives then Britain will need to find other mechanisms to deliver the obligations it has accepted under international treaties. There may be a variety of responses to these administrative changes; differences between the nations of Britain may result in further devolution of the administration of nature conservation. There has never been a time of such uncertainty in how the government manages our relationship with wild nature. As in the debates of the early 1940s, there is all to play for.

Resurgent utility

Through the 1990s debates at international conventions started to consider the interdependency of humans with the natural world. There were growing anxieties that ill-considered economic growth was undermining the ability of the planet to sustain people; environmental issues were now much broader than biologists lamenting the catastrophic losses of biological diversity. In response to these concerns, theoretical concepts of ecosystem services began to find favour amongst policy makers. In 2000 the Secretary-General of the United Nations commissioned the Millennium Ecosystem Assessment through which international collaborators considered not only our obligations to the natural world but also our dependency on the services it provides us.

Army live-firing ranges at Fingringhoe, Essex. One of the many ecosystem services provided by saltmarshes.

The Millennium Ecosystem Assessment published by the United Nations in 2005 prompted the British Government to produce a United Kingdom National Ecosystem Assessment. In 2014 the team undertaking the assessment considered the role of saltmarshes in capturing and storing carbon as part of the nations' response to the challenges of climate change. This assessment recognised that different saltmarshes varied in their potential for carbon capture, with the muddy Solent sequestering 2.19 tonnes of carbon per hectare per year compared with the sandier substrates of North Norfolk capturing 0.64 tonnes. As there is a world market in carbon sequestration, a cash value can be put on this ecosystem service and so it can be included in Treasury calculations.

Another value of saltmarshes was explored when the Natural Capital Committee's 2015 report to the Westminster parliament's Economic Affairs Committee cited the case of Medmerry, a realignment of sea defences in West Sussex. The recurring annual cost of maintaining the original sea defences averaged out at £300,000, a cost that did not reduce the risk for the following season. An investment of £28m by the Environment Agency involved remodelling 450ha of the coast including building seawalls to protect local communities from the risk

The stump of a redundant seawall, Medmerry.

of flooding. Some 180ha of the lowest ground was returned to its historical condition as a saltmarsh, with allowances for that marsh to migrate as sea levels rise. The revitalised saltmarshes contribute to the sea defences in absorbing wave energy before it reaches the seawalls. The benefits accruing from the investment were calculated at £90m, a sound return on public investment. At Medmerry the valuations and consequential decisions favoured wildlife; that will not always be the case. There is a risk in uncritically allying nature conservation with simple economics.

Following a century of saltmarshes being defined by ecologists there is a resurgence in much earlier definitions founded on their utility. The ecosystem services provided by saltmarshes in Britain are diverse. Such thinking provides an opportunity to recognise the importance of saltmarshes that goes beyond a philosophical respect for the natural world.

Ecosystem services provided by British saltmarshes

Amenity
Biodiversity
Carbon sequestration
Conversion of intertidal into terrestrial land uses (historical)
Flood defence through storm and erosion buffering
Gathering of wild foodstuffs, including wildfowling
Gathering thatching materials (historical)
Gathering wild medicinal herbs (historical)
Health, a place for fresh air and exercise
Horticultural turf
Livestock farming
Manufacture of chemicals (historical)
Military defence and training
Nursery grounds for commercial fish/shellfish
Nutrient and sediment storage
Purification and filtration of water
Tourism
Waste disposal (historical)

Invasive and non-native species

chapter
eighteen

Worldwide, one of the greatest threats to wildlife is the rise of invasive non-native species. British saltmarshes are not immune to such risks but they are at an equally, arguably greater, risk from being impoverished through the dominance of a handful of native species.

Plants

There are remarkably few non-native plant species invading the fully saline habitats of British saltmarshes. This is predominantly due to the wide distribution of species associated with saltmarsh conditions; if a plant can tolerate intertidal habitats in our latitudes then it is likely already to be native.

Setting aside the debate on the antecedents of Common Cord-grass *Spartina anglica*, the most regularly occurring non-native plant of saltmarshes is Buttonweed *Cotula coronopifolia*, a highly desirable addition to the margins of garden ponds. Assisted by the horticultural trade, this native of South Africa has colonised much of the temperate world including southern England. As a short-lived perennial, Buttonweed does particularly well where paths, hoof prints and goose-grounds provide a continuous supply of bare patches for germination. I have known the plant in a Solent saltmarsh for over 20 years where it appears to be settled into a niche without apparently displacing native species. Neil Sanderson (2008) described it as having formed a novel but consistent plant community where feral Canada Geese *Branta canadensis* maintain a close-grazed sward of Sea Club-rush *Bolboscheonus maritimus* and Common Reed *Phragmites australis* where

OPPOSITE PAGE:
Skunk Cabbage in
estuarine alder wood,
Hampshire.

Buttonweed grows alongside Brookweed *Samolus valerandi* and Celery-leaved Buttercups *Ranunculus scleratus*. Canada Geese are non-native and arguably invasive; however, their presence can be beneficial to the suite of species dependent on short turf. The combination of cattle and Canada Geese is used by nature reserve managers to fine-tune the height of saltmarsh turf ready for the arrival of wintering flocks of grazing duck.

There is a danger in complacency when considering the invasive qualities of non-native plants. Most of these species are benign, their presence being little more than a curiosity. However, there is a small percentage of non-native species that, once established, go through a phase of rapid expansion causing loss of quality, sometimes to the point of destruction, of valued native habitats. The difficulty facing practitioners is separating out the curiosities from the latent threats. If Buttonweed proves to be a problem in the future our successors may rightly look unkindly on our inaction during the early years of its establishment.

A wider range of invasive non-native plants may be found in the brackish reaches of saltmarshes and include highly problematic species. New Zealand Pigmyweed *Crassula helmsii* develops continuous swards, dominating watercourses and seasonally flooded hollows in grazing marshes. In such circumstances it excludes native species, notably aquatic species and the annuals of muddy hollows including goosefoots and oraches, *Chenopodium* and *Atriplex* spp. Experiments in freshwater wetlands have yet to find an effective means of controlling *Crassula*. Total immersion in full-strength seawater for prolonged periods appears to kill the plant but that technique is equally effective in suppressing all vegetation. Unfortunately, where this has been tried *Crassula* has been known to recolonise as the native vegetation recovers. New Zealand Pigmyweed is a species we may have to learn to live with; grazing cattle over stands on less productive soils can suppress its dominance, so until someone comes up with something better, that is the best we can hope to achieve.

Brackish waters are readily colonised by Floating Pennywort *Hydrocotyle ranunculoides* and the floating Water Fern *Azolla filiculoides*, both of which have the capability to completely smother grazing marsh ditches. In the modern age *Azolla* is native to the tropical Americas but in previous interglacials Godwin (1975) demonstrates it was widely distributed across Europe with its spores being recovered from ancient sediments in both Britain and Éire.

American Skunk-cabbages *Lysichiton americanus* are found in one of the estuarine alder woods of the New Forest where they have proved remarkably resilient to increasing salinity as tidal influences migrate upstream. Having become established in the 1980s, the monumental leaves of this strikingly beautiful aroid spread to dominate an estuarine fen to the exclusion of most other herbaceous species. What started as a botanical oddity ended up as a threat. Fortunately skunk-cabbages respond to herbicide treatment and are currently in the process of being eradicated.

Occasionally, one encounters other non-natives in saltmarshes. Garden-origin hybrids of north American Michaelmas-daisies *Aster* spp. can become locally dominant, particularly in brackish reaches. Tamarisk *Tamarix gallica*, a stalwart shrub of coastal gardens, colonises saltmarshes through suckering but has yet to acquire a vigour that excludes native species. Other exotica of saline habitats provide potential sources of future colonists; current candidates include Summer-cypress *Bassia scoparia*, Saltmarsh Aster *Aster squamatus*, Shrubby Orache *Atriplex halimus* and Tree Groundsel *Baccharis halimifolia*, all of which have a proven ability to thrive in salty conditions and are currently present in terrestrial habitats close to saltmarshes. It is possible the Saltmeadow Cord-grass *Spartina patens* of Chichester Harbour is from garden origin. The demand for horticultural novelties offers endless opportunities for the introduction of non-natives that may go on to become invasives.

Invertebrates

Estuaries and tidal rivers, with their waters of varying salinity, are amongst Britain's most vulnerable habitats to invasion. Through inadvertent and deliberate means a diverse invertebrate fauna has been introduced into our coastal waters. American Hard-shelled Clams *Mercenaria mercenaria* are a commercially important shellfish and have become established in intertidal mud amongst the saltmarshes of the Solent. The impact of this colonist on the native infauna is unclear, but its presence is catastrophic when it comes to attempts to harvest it by dredging through eelgrass beds *Zostera* spp. Regulation and enforcement are now necessary to prevent destruction of these habitats which are of importance to native commercial species, not least Bass *Dicentrarchus labrax* nurseries and some of the last remaining native Common Oyster *Ostrea edulis* beds of the region. Hard-shelled

Clams are just one of a suite of non-native bivalves spreading through Britain's tidal rivers, with Zebra Mussels *Dreissena polymorpha* colonising as far north as central Scotland and the newly arrived Quagga Mussel *Dreissena rosteriformis bugensis* threatening to do likewise.

Another major group of invertebrates of tidal waters is the Crustacea, which include crabs, crayfish and shrimps. Chinese Mitten Crabs *Eriocheir sinensis* were first reported in southern England in the 1930s, since when they have colonised tidal rivers as far north as Cumbria and Northumbria. For most of their lives Mitten Crabs live in freshwater, migrating in autumn to breed in tidal waters where they burrow into the face of saltmarshes and mudflats for shelter. Once hatched, the larvae develop within the sheltered waters of the marsh before returning upstream as juveniles to live in the sediments of soft riverbanks. This is the only freshwater crab in Britain and is unmistakable with its furry mitten-like claws.

In east Asia, Mitten Crabs are a highly prized seafood and it may be that they become a commercially important species in Britain. Booy *et al.* (2015) provide useful accounts of culinary introductions of brackish-water tolerant crayfish into Britain since the 1970s. Retrospective legal action is too late to prevent the establishment of Red Swamp Crayfish *Procambarus clarkii* from the southern United States and Mexico, Narrow-clawed Crayfish *Astacus leptodactylus* from the Black Sea, and Spiny-cheek Crayfish *Orconectes limosus* from the American Atlantic seaboard.

Chinese Mitten Crabs.

Smaller still are shrimps which, like some crayfish, can exhibit broad environmental tolerances. Since the 1930s the Northern River Crangonyctid *Crangonyx pseudogracilis* has extensively colonised British fresh and brackish waters south of the Highland line. The Bloody-red Mysid *Hemimysis anomala* is a relative newcomer from the Black Sea, being first recorded in Britain in 2004 but making good progress through the English Midlands. In Europe this mysid has developed strongholds in ports and harbours and so may be expected to colonise the less exposed reaches of our intertidal coast. Demonised by nomenclature is the Killer Shrimp *Dikerogammarus villosus*, another native of the Black Sea, being first recorded in Britain in 2010, which is currently expanding through the weakly brackish wetlands of the Norfolk Broads.

It remains unclear what impact these colonists may have on our native wildlife. The places they first become established tend to be amongst highly modified habitats and so potential impacts are difficult to assess. Interactions of non-native species with established ecosystems are complex, with impacts ranging through subtle modifications of local natural history to threats to whole ecosystems including human life and commerce. With so many factors to consider, none of which remains constant, it is practically impossible to quantify objectively such impacts until it is too late to take remedial action. Conservation effort is therefore currently directed to preventing the arrival of additional species, stopping the spread and, where possible, eradication.

Red Swamp Crayfish.

Mammals

Of all the non-native species of British saltmarshes, Domestic Sheep *Ovis aries* are so familiar that we have adopted them as an honorary native. Domesticated sheep are descended from the wild Urial *Ovis orientalis* whose native range runs through the mountains of western Asia. The earliest evidence of domestication comes from about 10,800 years ago in modern-day Iran. Sheep and shepherding gradually spread across Europe, a progression associated with a shift in human culture from hunting wild animals to herding domestic stock. Sheep farming crossed the Alps by 6,500 years ago, after which it took another thousand years to arrive in Britain. The earliest archaeological evidence of sheep in our islands is from Berkshire. Despite this first record being from the south's chalk country, Derek Yalden (1999) interprets evidence from the New Stone Age as showing sheep were much more important in the economy of the Northern and Western Isles than in the lowlands. By the time metal-working technology became widely adopted, some two thousand years later, cattle and sheep had become the principal livestock of British agriculture.

Sheep are non-native but cannot be categorised as invasive. The wet climate of Europe's Atlantic margins is far from ideal for a species that evolved in the hot, dry conditions of the Near East. The primary job of a shepherd is to keep their flock alive. Over the last five millennia a combination of breeding, chance and local evolutionary pressures has generated native breeds of sheep displaying characteristics adapted to the British landscape. The long history of keeping sheep in the challenging climate of the Northern Isles helps explain the robust nature of northern short-tailed breeds with their tolerance of foul weather and a penchant for coastal forage ranging from saltmarshes to seaweeds. One of the advantages of evolving in an arid climate is a limited requirement for freshwater, which supports the exploitation of coastal habitats well away from springs and rivers.

Whereas domesticated cattle and equines are ecological proxies for their wild forebears, there is no such continuity relating to sheep; for most of ecological history British saltmarsh habitats evolved in their absence. In considering the place of sheep in British ecosystems it is recognised that they share some behavioural traits with equines, as the method of grazing by both species is a scissor-like bite, allowing the turf to be cropped closely. Both species are communal, individual animals finding safety in numbers and so tending to concentrate grazing pressures in discrete areas at any one time. Similarly both

Shoreline grazing on
North Ronaldsay, Orkney.

species tend not to drop their dung at random; equines set aside latrine areas and sheep hold their dung until nightfall. These traits can still be witnessed in the behaviour of 'primitive' breeds when kept unherded on extensive grazings.

Fallow Deer *Dama dama* are another non-native mammal of the lowlands. This species of southern Europe was introduced into Britain during the Roman occupation, the initial herds being kept as parkland animals. Analysis by Madgwick *et al.* (2013) of radio isotopes in bones excavated at the Flavian palace near modern Chichester in West Sussex suggests the herd was grazed on the adjacent saltmarsh. Conditions in Roman Britain were however unsuitable for the deer outside the protective confines of a park. They died out but were reintroduced by the Normans, eventually spreading from parks and royal forests into the countryside. Fallow Deer feed by both browsing and grazing and their droppings can be found on many saltmarshes, particularly those close to woodlands or other undisturbed habitats. In my experience the grazing pressure exerted by Fallow Deer exploits, but does not modify, the vegetation.

In contrast Hannaford *et al.* (2006) have shown how Sika Deer *Cervus nippon* grazing can change cord-grass *Spartina* spp. marshes to such a degree that they become dominated by finer swards of saltmarsh-grass *Puccinellia* spp. and glassworts *Salicornia* spp. This appears to be a localised effect, being conspicuous in Poole Harbour, Dorset, where Sika were introduced in 1896, but not on the Beaulieu River, Hampshire, where they escaped from captivity in 1904. The different impacts are likely to relate to the degree of stalking on the respective estates. Since their introduction in the late 19th century to numerous sites across the British Isles, Sika have spread to colonise substantial parts of the afforested Highlands with numerous outlying populations in England. Sika are sufficiently closely related to Red Deer *Cervus elaphus* to hybridise with them so that many, if not most, populations contain some introgression. By grazing saltmarshes Sika are proxies for the native Red Deer and are an important factor in maintaining open habitats in otherwise ungrazed coastal wetlands.

Not all of our native mammals are native across the islands that comprise Britain. Red Deer were absent from the Outer Hebrides and Orkney until being introduced by New Stone Age farmers around 5,000 years ago. Genetic studies by Stanton *et al.* (2016) have shown these were not animals transported from the nearby mainland but from an unknown location much further afield than either the British mainland or neighbouring Europe. Red Deer and sheep are important components of the saltmarsh ecosystems of the Highlands and Islands; in the case of the Outer Islands neither species is native.

Sika graze the saltmarshes of Poole Harbour, Dorset.

Native invasive species

In the absence of native herbivores or their proxies many types of saltmarsh vegetation are vulnerable to being dominated by a limited range of coarse perennial species. Specialist species of saltmarshes, be they plants, invertebrates or birds, are mostly associated with structurally complex marshes. That structural diversity is best developed where local topography is enhanced by selective grazing.

The simplification of species composition and structure in the absence of large herbivores varies depending on the place of a saltmarsh in the tidal range. Freshwater transitions are vulnerable to domination by Common Reed *Phragmites australis* and occasionally Alder *Alnus glutinosa*. In more frequently flooded upper saltmarshes Sea Couch *Elytrigia atherica* can become dominant, as can finer grasses which have the potential to generate lank mats of Red Fescue *Festuca rubra*, Creeping Bent *Agrostis stolonifera* and Common Saltmarsh-grass *Puccinellia maritima*. These communities in turn grade into lower marshes with a tendency to develop near-monocultures of Sea-purslane *Atriplex portulacoides*. Historically, the foreshore alone was immune to dominant perennial species until the emergence of Common Cord-grass *Spartina anglica* in the 19th century.

It is competition from vigorous perennials that is a common threat to most of our rarest native species. Cheffings and Farrell's 2006 Red Data Book of flowering plants identifies two saltmarsh plants as critically endangered and four more as endangered. Both of the critically endangered species, Pedunculate Sea-purslane *Atriplex pedunculata* and Triangular Club-rush *Schoenoplectus triqueter*, are at immediate risk of functional extinction in the wild, their retention in our native flora only being secured through continuous management interventions including reinforcement from seed banks. Both species need open conditions that historically were maintained by livestock. Similarly, all four species on the endangered list are associated with short open habitats; being Least Lettuce *Lactuca saligna*, Shore Dock *Rumex rupestris*, Small Cord-grass *Spartina maritima* and Curved Sedge *Carex maritima*. Least Lettuce is the most vulnerable to the lack of livestock as every year it needs bare ground in which to germinate. The others are perennials with a degree of resilience to short-term changes but only the itinerant Shore Dock is likely to persist in the prolonged absence of large grazing animals. As the threat levels reported in the Red Data Book decrease, so the number of species in each category increases. The pattern of threat remains the same; less vigorous native species are overwhelmed by their robust compatriots where saltmarsh ecosystems have been disrupted through the exclusion of large herbivores.

The invertebrate fauna of saltmarshes is much more diverse than the flora but is less well understood. Close inspection will reveal that every species of plant will have its invertebrate associates, from the dominant invasives to the most ephemeral rarity. As with flowering plants, the richest saltmarshes for invertebrates tend to be large sites with a high degree of structural diversity, particularly where freshwater is present. There are many accounts of the entomology of saltmarshes that reflect a cultural aversion by some naturalists to 'overgrazing'. There are frequently expressed anxieties that grazing has been responsible for the destruction of important invertebrate foodplants, notably Sea Aster *Aster tripolium*, sea-lavenders *Limonium* spp. and Sea Wormwood *Artemisia maritima*. There may well have been such occasions of loss but in practice all three species prosper under a variety of extensive grazing regimes. Indeed, taking the long view, grazing is essential in enabling the continuity of populations of these plants as saltmarshes migrate landward. Herbaceous perennials may be persistent once established but they need open ground to germinate and form new populations; they cannot achieve this through a thatch of rank grasses.

To a botanist the inclusion of Sea Wormwood amongst the vulnerable invertebrate foodplants is unexpected as it is an exceptionally resilient species capable of thriving even under sheep grazing, that most defoliating and potentially harmful of practices. Old leggy specimens of Wormwood are vulnerable to being broken up by trampling hooves; the risk being greatest when grazing pressures are suddenly intensified. Sea Wormwood is however a classic plant of grazed environments, being endowed with bitter anti-feeding compounds, an effective evolutionary response to defoliation, and it is an exceptional host plant for a diverse invertebrate fauna.

Invertebrates that are dependent on specific plants during critical stages of their life cycles are vulnerable to inappropriate grazing regimes. For example, whilst Sea Aster will happily persist in a close-grazed sward in a non-flowering state, this vegetative condition will be fatal to invertebrates that specialise in eating its pollen. Competition for the same resource, in this example flowers, does not need to mean that the competitors are mutually exclusive. The diverse species of saltmarsh ecosystems co-evolved over millennia and, until recently, grazing by large herbivores was the normal state of affairs on most marshes. However, given the perilous state of some invertebrates, it would be wise to invest in understanding the ecological relationships between the species of any particular marsh before either reinstating, or confirming the exclusion of, large herbivores.

There are real risks to biological diversity in overgrazing saltmarshes but there are other, arguably greater, risks in abandonment. In 2004 I wrote a review of the effects of different grazing regimes by domesticated animals on British saltmarshes. This review

The bee *Colletes halophilus* is dependent on the continuity of tall flowery stands of Sea Aster.

considered the different ecological effects arising from grazing various domesticated and native animals at a range of stocking densities. I found that the greatest diversity of habitat structure was generated by domesticated proxies of native cattle and equines grazing large saltmarsh landscapes at low densities. High stocking rates of sheep and cattle have the potential to simplify the structure of a marsh through the processes of defoliation and trampling; however, short grassy saltmarshes have important associated species that contribute to natural diversity. Artificially high grazing pressures, sustained through supplementary feeding or seasonal mob-grazing, can be devastating to diversity to a degree that even resilient species may succumb; but even then there are specialists of extreme conditions, the denizens of muck and mud. A large and complex enough site should have room for all aspects of the ecosystem to be expressed, from the poached gateway to the isolated creek-marsh. The presence of large herbivores on saltmarshes is entirely natural and an integral part of saltmarsh ecosystems; the challenge to conservationists is how to mimic these natural ecological processes in an unnaturally fragmented countryside. Conservation management of saltmarshes would improve if practitioners considered what degree and character of grazing was appropriate rather than take ideological positions on the equally devastating extremes of all or nothing.

Conservation

in practice

L aw and policy set out the framework within which practical people deliver nature conservation. Most conservation initiatives can be placed somewhere on the continuum from saving individual species to securing the ecological processes necessary to safeguard suites of habitats. The following accounts draw on my own experience in the south of England to explore that continuum. I liken single species initiatives to the role of an intensive care ward; the patient is kept alive through a time of crisis with the aspiration of returning them to an independent life. Such emergency measures are sometimes necessary, but most wildlife lives outside such close management regimes and needs a different approach.

Fisher's Estuarine Moth

In the autumn of 1968 Ben Fisher noticed a large and conspicuously colourful moth on the lighted window of his home at Beaumont-cum-Moze. Fisher was a skilled naturalist. He knew that this was something special and so he sought out a name. The moth was a noctuid, broad bodied with its upper wings banded through various honeyed tones, punctuated with pale shields and cryptic emblems. Superficially it looked like a Frosted Orange *Gortyna flavago* but it was far too large.

The specimen was subsequently identified as *Gortyna borelii*, new to Britain, and rather than adopting one of its continental synonyms it was granted the English name of Fisher's Estuarine Moth. There is an honourable tradition in Britain of naming moths after their discoverers. The genus *Gortyna* has been recognised since 1816 and the species *borelii* since 1837. In Britain Fisher's Estuarine Moth is at

OPPOSITE PAGE:
Titchwell Marsh, Norfolk.

Fisher's Estuarine Moth.

the very edge of its European range from where it can be found in a variety of continental and coastal habitats between the Black Sea and the Atlantic.

Ben Fisher's home was on the edge of Hamford Water, an embayment of the outer Thames estuary sheltered from the North Sea by the cliffs of the Naze. Hamford Water contains a maze of creek marshes and low islands which throughout the recent historical past have been farmed and cropped for wildfowl. Through the 19th and early 20th centuries the Ordnance Survey mapped the marsh islands as conventional fields without requiring the symbols of rough grassland or scrub. By the interwar years Dudley Stamp's land utilisation survey (1942) regarded most of them as rough pastures of various categories with only the largest island, Horsey, sustaining arable crops. By the 1960s agriculture of any kind was in retreat on all islands except Horsey, so by the time Fisher found his moth many of the marsh islands had developed a cover of tall grasses and scrub.

Investigations by Colin Hart (1999) established that the moth was confined to a few sites around Hamford Water. Various survey techniques were available to estimate the size of the populations but one of the best was looking for the frass excreted by the caterpillar as it fed inside the stems of Hog's Fennel *Peucedanum officinale*. Zoe Ringwood *et al.* (2002 & 2004) has described how the adult moth lays her eggs on coarse grasses close by the foodplant. Upon hatching, the caterpillar bores into the higher parts of Hog's Fennel stems, gradually moving down inside the plant until taking up residence in the root from where it burrows out to pupate in the adjacent soil. The caterpillar needs to have a strong constitution; Hog's Fennel is highly aromatic, containing complex chemical compounds to protect it from herbivores, both large and small. The 17th-century herbalist Gerard referred to the plant as Sulphurwort, observing that the root gave out a grievous smell with the viscous sap waxening hard when exposed to air. Any caterpillar that can thrive whilst excavating through such a hostile host is unlikely to encounter casual predators. Fully grown caterpillars are at greatest danger from one another; cannibalistic tendencies mean that only one adult usually emerges from any single rootstock.

Hog's Fennel is strongly associated with the Thames estuary and historically has enjoyed a scattered distribution across the grazing marshes of both shores. As a valuable plant in the apothecary's trade it is amongst the earliest species of London's marshes to have been

Seedlings of Hog's Fennel establish freely in broken ground.

recorded by botanists, and as a national rarity detailed recording effort has been sustained to the present day. Fisher's Estuarine Moth requires coarse grasses in which to lay its eggs whilst Hog's Fennel requires bare ground in which its seeds may germinate. A species conservation programme for the rare plant is likely to have different management priorities than for the rare moth.

The ecological requirements of the plant and the moth are not as contradictory as may first appear. Hog's Fennel is a long-lived plant reproducing vegetatively from root divisions as well as from seed. Nevertheless, every so often it needs bare ground in which new plants may germinate, not least when populations need to shift to keep pace with sea-level rise. The islands of Hamford Water are no longer in a state where Hog's Fennel can readily reproduce itself; sitting amongst its fields of scrub and rank grasses, it is confined to an ecological backwater.

Returning to the moth, the risks to its survival have been identified as inappropriate management along with sea-level rise, inappropriate management in this case being the re-establishment of historically recent grazing regimes that may suppress the coarse grasses in which eggs are laid. Sea-level rise is a very real threat, with the prognosis for the survival of the marsh islands being bleak. To conserve the moth it was therefore decided to take Hog's Fennel into cultivation and to introduce it to sites along the Essex coast. To date, over 25 plots have been bedded out with tens of thousands of seedlings. The locations of these sites include those at elevations well above Hog's Fennel's natural place in the tidal range. Like many herbaceous perennials, once well established a young plant of Hog's Fennel will thrive where planted, be it a garden or on the gravel plateau of a country park. The natural ecological range and distribution of this botanical rarity are now masked by numerous introductions. The favoured moth is far from sedentary and unaided it has colonised the majority of these plantations irrespective of which habitats it was introduced into. Should it require further assistance the moth is being bred in captivity and has been successfully translocated to Hog's Fennel planted far from its origins in Hamford Water; this is single species conservation on a landscape scale.

There is insufficient evidence to determine whether Fisher's Estuarine Moth is a newcomer to Britain that is benefiting from the tall-grass swards that developed after the abandonment of grazing marshes. Similarly, there is not the evidence to tell whether the moth has a longer history persisting undetected in sub-optimal conditions whilst the marshes were managed as part of the local pastoral economy. This later sub-optimal state is the fate of most species in areas of high diversity; in our richest habitats few species thrive but many survive. By optimising a landscape for a single species you inevitably prejudice against overall diversity.

The drive to expand Britain's population of Fisher's Estuarine Moth is indisputably successful. What the initiative has created is a series of innovative habitats that are reliant on a continuous regime of conservation management. The venture is sustainable as long as conservationists continue to intervene. To return to the opening analogy, in Britain populations of Fisher's Estuarine Moth remain in the intensive care ward; in the meantime the moth has status, a high public profile and champions dedicated to its cause.

Scrapes and reedbeds

The promotion of Fisher's Estuarine Moth represents a single species initiative. Elsewhere, saltmarsh nature reserves have been developed in other ways to promote discrete suites of species, notably breeding birds associated with lagoons and coastal reedbeds. Saltmarshes are naturally dynamic, a dynamism that may be directed for good or ill through a combination of engineering and biological science.

Nature conservation is subject to changing perspectives and priorities. By the late 1960s conservation organisations were gaining the confidence and capital to acquire nature reserves and develop them to meet ambitious objectives. Through experience of sites such as Minsmere on the Suffolk coast, by the late 1960s there was a growing understanding of the ecological requirements of species such as Avocets *Recurvirostra avosetta*, Bitterns *Botaurus stellaris*, Bearded Tits *Panurus biamicus* and Marsh Harriers *Circus aeruginosus*. Saltmarshes, particularly at their freshwater transitions, are biddable habitats that can be manipulated, even transformed, for the benefit of these emblematic rarities.

This historical trend is exemplified in the RSPB's Titchwell Nature Reserve on the North Norfolk Coast. The storm surge of 1953 breached Titchwell's seawalls so that when the RSPB purchased the site in 1973 the habitats comprised a complex of early-stage succession saltmarshes and tidal reed. It took the rest of the decade to repair the walls and install dams and sluices, which transformed the saltmarshes

Bearded Tits readily colonise newly made reedbeds.

into freshwater reedbeds and other wetlands of varying salinities. The precise contours of the marshes were tailored with knowledge of the micro-habitat preferences of target species; the colloquial term of 'scrapes' for such precision engineering belies a highly advanced understanding of practical ecology. The diversity of breeding birds at Titchwell rose from 39 species in 1975 to 61 in 1982 in response to intricate habitat management. Visitor numbers similarly increased from 500 a year in 1973 to 69,000 in 1989; today numbers are closer to 100,000. The work was entirely successful in meeting the management objectives set in 1974 to increase and diversify breeding and migratory birds, together with engaging the public; the cost of this achievement was £70,000 and the loss of 36ha of saltmarsh.

In the late 1980s the RSPB reviewed their management of Titchwell. In a refreshingly candid study they considered the perception of saltmarshes that prevailed at the time they acquired the site and compared it with current thinking. The early-stage succession saltmarsh communities that had developed following the 1953 breaches were regarded in the 1970s as of little intrinsic nature conservation importance. This perception changed during the 1980s with the growing understanding of the importance of such intertidal habitats in a European and global context. The engineering had created a bird reserve of national and international importance but at the expense of equally important habitats.

In a section of the review by Becker & Stills (1988) titled 'Death of a Saltmarsh' there is an account of how saltmarsh communities of the lower and middle marsh at Titchwell were suppressed by freshwater flooding. From 1976 the species composition of the engineered sections was monitored, revealing that glassworts *Salicornia* spp., Sea Aster *Aster tripolium* and sea-lavenders *Limonium* spp. went into rapid decline, with only Common Saltmarsh-grass *Puccinellia maritima* proving slightly more persistent. There was a temporary boom in Red Goosefoot *Chenopodium rubrum* which was regarded as a problem until Avocets were found to appreciate the cover it provided. In time the stands of goosefoot died away as the face of the Titchwell's reedbeds colonised the former intertidal at over a metre a year.

The conclusions of the review addressed both practical and conservation issues:

> *Titchwell exemplifies a pattern where management leads to more*
> *management … Anyone interested in the conservation success of*
> *the marshes at Titchwell must also be realistic about the cost*
> *of maintenance.*

The final paragraph concluded:

> *A decision to convert a saltmarsh to a non-tidal marsh should, however,*
> *always be based on a sound knowledge of the ecological importance of*
> *the saltmarsh concerned, in both a national and international framework.*
> *Where a saltmarsh is demonstrated to have inherent conservation*
> *importance, then it should be safeguarded in its own right, and conversion*
> *to habitats similar to those at Titchwell should be resisted.*

The most recent round of sea defence works at Titchwell has moved towards returning habitats created in the 1970s to the intertidal. The importance of presenting wetland birds to the public is unquestionable but those laudable aims need not be achieved at the expense of other valuable habitats. In recent decades the RSPB and other conservation organisations have been wise in investing in the creation of freshwater wetlands and associated visitor facilities in locations well above the immediate risks of sea-level rise.

There remains a legacy of developments similar to Titchwell in saltmarshes managed by various organisations throughout Britain. The storm surge of 2013 resulted in many nature reserves along the East Anglian coast becoming temporarily overwhelmed by saltwater. The emotional rhetoric in the media following the floods suggested that wildlife had been devastated, stories that were then repackaged as nature naturally recovered. There was costly damage to hides and other visitor infrastructure, a matter that was quickly addressed with financial support from government agencies. Visiting bird reserves is now an important part of the rural economy. Elsewhere on the East Anglian coast there were some significant changes arising from the 2013 surge. The damage to sea defences at the Suffolk Wildlife Trust's Hazelwood Marshes proved irreparable and so former grazing marshes have returned to the intertidal. At Hazelwood these uncalled-for changes have been embraced and the rejuvenation of this reach of the Alde estuary is now regarded as the new norm. Coastal nature reserve managers know that the next surge is not so much a question of 'If' as one of 'When'.

Fashions have their devotees, even for the excesses of the 1970s. There are still those who see estuaries as malleable habitats offering opportunities for enhancement of bird diversity and public enjoyment

Tasselweed pools in the Axe wetlands can be enjoyed by tram.

as an alternative to their restoration to wild nature. An upper reach of the Axe estuary on the Devon–Dorset borders has recently been comprehensively engineered to such ends with the accompanying paraphernalia of hides, blinds and boardwalks. Here at Colyford Common upper saltmarsh transitions, including internationally important tasselweed *Ruppia* spp. pools, are succumbing to the tyranny of reeds as livestock are excluded from the core of the marsh. Given enough time, the sea will overwhelm these adornments.

Brent Geese

Every autumn flocks of Dark-bellied Brent Geese *Branta bernicla* ssp *bernicla* make their way from breeding grounds high in the Russian Arctic to overwinter in the comparatively mild climate of Europe's Atlantic seaboard.

Dark-bellied Brent Geese came close to extinction in the middle decades of the 20th century. The primary cause of their decline was hunting throughout their wintering range. Brent Geese are about the size of a mallard and are reportedly good eating; they have long been the wildfowlers' quarry of choice. Before wildfowling developed its modern lethal efficiency Brent Goose populations were determined

In extreme weather Brent Geese resort to Farlington.

by conditions on their breeding grounds; with plentiful food and nest sites their breeding success is constrained by predation. The principal predators of Brent Geese are Arctic Foxes *Alopex lagopus*, which will take chicks when their preferred prey of Arctic Lemmings *Dicrostonyx torquatus* and Siberian Brown Lemmings *Lemmus sibiricus* are unavailable. In a year with many lemmings, the Brent Goose population may increase by over 30 per cent, but in poor lemming years nearly all the chicks are eaten. Through the early 20th century the combination of natural predation and human exploitation meant that Brent Goose populations became unsustainable.

Extinction was averted by the imposition of restrictions on shooting the geese throughout their European range, the ban coming into force in Britain in 1954. The world population built steadily over the succeeding decades from around 22,000 birds in the early 1960s to over a quarter of a million by the late 1990s. As a species with intermittent breeding success the population varies but the current long-term trend is still a slight increase. The world population is now estimated to be in the order of 300,000 birds.

The Solent supports the most important concentrations of Dark-bellied Brent Geese sites in Britain. In 2015 the combined harbours and estuaries were the wintering grounds of nearly 35,000 birds, a substantial proportion of the British population of a little over 113,000. The 1954 Protection of Birds Act safeguarded Brent Geese from persecution whilst the designation of Sites of Special Scientific Interest (SSSI), Ramsar sites and Special Protection Areas has sought to conserve the habitats on which they are dependent.

Early observers considered the Solent's Brent Geese as extreme specialists, feeding on little else but eelgrass *Zostera* spp. and algae. The spread of *Spartina* marshes across eelgrass beds, combined with a wasting disease of *Zostera* throughout the 1930s, appeared to explain the local decline in the geese. It now appears that in the early 20th century this behaviour was a response to wildfowling, the birds seeking out the safest feeding grounds and roosts. There are many accounts of geese resorting to roosting out at sea with flocks rafting-up to escape the attention of the guns. With the effective application of a shooting ban their behaviour changed.

Annual Beard-grass has been known from Farlington Marshes since 1595.

Colin Tubbs (1999) recounts how on 20 December 1969 a flock of 100 geese, mostly juveniles, moved onto grazing marshes by Langstone Harbour. The site was Farlington Marshes, which was one of the few places on the harbour undisturbed by wildfowling, having been secured as a nature reserve by the Wildlife Trust in 1962. Farlington Marshes as we know them today were created between 1769 and 1773 when the lord of the manor embanked 120ha of a saltmarsh peninsula. The flood defences have never been particularly efficient, there still being occasional overtopping at high tides and persistent leaks through the tide gates. Having been managed as summer pastures for centuries, the grasslands have retained much of their saltmarsh character, supporting all four native species of saltmarsh-grasses *Puccinellia* spp., together with Sea Clover *Trifolium squamosum*, Slender Hare's-ear *Bupleurum tenuissimum* and the Annual Beard-grass *Polypogon monspeliensis*.

Grazing the brackish margins of Shuts Lake, Farlington.

Brent Geese are gregarious birds living in extended family groups. The movement of a flock of juveniles away from established feeding grounds is inevitable if the population is rising beyond the immediately available stock of food. The breakaway group of 1969 would have found Farlington's grasslands to their liking; they had fed on similar vegetation in their Russian breeding grounds. Arrowgrasses *Triglochin* spp. and saltmarsh-grasses *Puccinellia* spp. do not change in palatability with latitude. Brent Geese like short grass and long horizons, the scale of the marshes providing open country to survey for the approach of predators. Summer grazing by cattle leaves a compact turf of tillering grasses, ideal grazing conditions for a small goose. Farlington had other advantages, not least it was free from regular inundation at high tide; here was a refuge and feeding ground. The juveniles that took to Farlington in December 1969 grazed the marsh until the following March.

The birds did not reappear in the following three winters, but by the mid-1970s the flocks had extended their range from the permanent grasslands of Farlington into adjacent playing fields. By the late 1970s they developed a taste for autumn-sown cereal crops. Conserving flocks of geese that had taken to use such places required a new approach. Sites of Special Scientific Interest are usually notified on the basis of selecting sites that support important habitats. By the time the Solent SSSIs were notified, most of the grassy upper saltmarshes of the region had been modified to the point where they would not meet SSSI selection criteria.

Embanked marshes of a similar character to Farlington were once widespread around the Solent, both in the sheltered harbours and along the more open shores to the west. Colin Tubbs calculated that 3,655ha of the Solent's intertidal has been modified to create grazing marshes. That resource has declined as those marshes that were not subsequently built over have been intensively farmed so that natural vegetation only survives in nature reserves or under protective designations. The spreading cities have overgrown their neighbouring coastland, the few remaining green spaces being converted to sports fields or cautiously cosseted lest disturbance exhume old landfill. Brent Geese do not make distinctions between quality habitats and urban grasslands, nor do they recognise the different qualities of arable and saltmarsh; if the site is safe and food is available then they will use it. To many residents of Portsmouth the spectacle of a winter sky darkened by waterfowl is a seasonal joy helping to define the very nature of their communities; the Brent Geese of Langstone Harbour are the apogee of urban wildlife.

Brent Geese over Portsmouth.

There is inevitable competition for land where there is urban growth, particularly so where that expansion is constrained by statutory conservation designations. Development issues around the cities of the Solent are particularly difficult to address as the populations of wild birds may be protected but some of the land they depend on is not. A series of planning disputes ran through the 1980s and 1990s, many of which were resolved through public inquiry. The case with the highest public profile was a proposal by Portsmouth Football Club to relocate to a new stadium on an embanked saltmarsh adjacent to Farlington. The public inquiry of 1994 concluded with refusal of permission to build as, despite the land having no statutory conservation designation, the international obligations for safeguarding Brent Goose populations overrode the club's aspirations.

Having proved the conservation case through public inquiries, a collaborative approach was subsequently adopted to integrate international obligations with managing urban growth. Volunteer ornithologists have surveyed over 1,000 fields and urban green spaces adjacent to the Solent to determine their use by Brent Geese and other waterfowl. Debbie King's (2010) report of the combined results identified Brent Geese using more than 390 fields covering some 2,000ha. A significant proportion of this area, some 39 per cent, fell outside the statutory conservation framework. At the heart of the conservation strategy for the eastern Solent is Farlington, the largest of the coastal grazing marshes upon which the wintering birds rely for feeding, roosting and a foul weather refuge. The security of the international importance of Langstone Harbour is dependent on its maintenance. Farlington remains a summer fattening marsh for cattle as it is large enough to be attractive to commercial graziers. As south Hampshire becomes increasingly urban, and livestock farming declines in both presence and profitability, so that enterprise becomes increasingly marginal. Securing livestock and the skilled people who care for them is an annual issue but longer-term trends are also in play. Despite Farlington enjoying every conservation designation and being managed as a nature reserve, it remains a saltmarsh, modified through embankment, and someday it will return to the intertidal, if not by design then by default. At present the walls are subject to running repairs, a holding operation until such time as a decision is made. The wildlife of the marshes is of international importance, not just for the birds. If the wall breaches irreparably then those losses will need to be made good; the challenge is finding somewhere that wildlife can move to.

The Lower Test exemplum

Coastal saltmarshes pose a particular challenge to the concept of site protection as they do not stay in the same place for very long. All habitats and species are mobile, at least over evolutionary time, but saltmarshes move within the short lifespans of political administrations. A commitment to conserving saltmarshes requires an ability to be as flexible as those natural processes.

An intertidal zone, with its saltmarshes, will naturally migrate upstream in response to rising sea levels wherever the contours are amenable. Should the landward face of that zone be engineered with a seawall then that natural migration is artificially constrained and the zone becomes 'squeezed'. A squeezed intertidal zone will become narrower over time and will eventually be lost.

Coasts with the strongest expression of sea-level rise are in southern Britain where the major estuaries are either embanked, urbanised or exhibit the naturally rising ground of incised river valleys. A rare example of an un-engineered marsh can be found at the head of Southampton Water in the Hampshire and Isle of Wight Wildlife Trust's Lower Test Nature Reserve. The upper estuary of the River Test is unconstrained so that every high tide brings a merging of fresh and saltwater across the floodplain. That floodplain is free from significant engineering; the limits of the floods are defined by natural

Redbridge, where the tidal Test enters the estuary of Southampton Water.

terraces. The topography of the Lower Test Marshes is not pristine because a causeway partially crosses the valley where the anastomosing freshwater channels recombine as they enter the estuary. Fortunately the bridges at this point are sufficiently wide to provide free passage for the tide through and around that causeway.

The combination of these factors has resulted in the survival of a major lowland river making its sea fall through a complex of estuarine habitats. The reach of the valley subject to both tidal and freshwater influence is some 2km long and 1km wide, the marshes extending over 160ha. The vegetation of the Lower Test is highly diverse; surveys using the National Vegetation Classification have identified over 40 communities in that short stretch of floodplain. Neil Sanderson's survey of 2008 described 15 saltmarsh communities (including five sub-communities of Red Fescue *Festuca rubra* and Saltmarsh Rush *Juncus gerardii*), 13 types of swamp, six mesotrophic grasslands and a mire together with an assortment of open water, parched grasslands and other vegetation of open habitats.

In its seaward reaches Lower Test's saltmarsh is typified by fine grasslands with glassworts *Salicornia* spp. in muddy pans, and with Sea-purslane *Atriplex portulacoides* and Common Sea-lavender

Fattening pastures and Southampton docks.

Limonium vulgare along the better-drained creek edges. The grasslands hold a population of Bulbous Foxtail *Alopecurus bulbosus* together with its hybrid *Alopecurus* × *plettkei*. The sward is disturbed by both cattle and geese, creating conditions for ephemeral communities such as that characterised by Lesser Sea-spurrey *Spergularia marina* and Reflexed Saltmarsh-grass *Puccinellia distans*. In the Lower Test this vegetation is diversified by the presence of Stiff Saltmarsh-grass *Puccinellia rupestris* in what appears to be a southern species-rich variant of the community.

The 2008 survey was one of a series of studies dating from the mid-1990s through which it is possible to track the changing nature of the Lower Test marshes. The surveys were prompted by the death of willows *Salix* spp. which were once a prominent feature in the swamps. These willows had provided cover for breeding Cetti's Warblers *Cettia cetti* in the early years of their colonisation of Britain. A combination of stress from increasing salinity and fungal infections had proved fatal. That deadwood has been left to stand to be progressively embraced within the intertidal. Throughout this period the flow of the main river maintained predominantly freshwater habitats along the riparian margins. Saltmarshes and brackish swamps flanked that

Saltmarsh migration overtaking stands of willows.

Divided Sedge.

zone, their structure determined not only by the salinity but also by the distribution of cattle together with mowing of reedswamps. A decade proved long enough to record the response of this vegetation to sea-level rise.

Saltmarsh communities are colonising the swamps of the Lower Test; the establishment of species such as Wild Celery *Apium graveolens*, Divided Sedge *Carex divisa* and Saltmarsh Rush *Juncus gerardii* is assisted by the opening up of the swards by cattle, which most frequently occurs in brackish parts of the Greater Pond-sedge *Carex riparia*, Reed Sweet-grass *Glyceria maxima* communities. The most resistant of the tall grass communities to change are the reedswamps, but they too become diversified in the presence of grazing, the cattle creating a matrix of paths and saltmarsh glades.

Livestock have unrestricted access to the tidal river so they intermittently trample and graze the intertidal flats. These flushed briny conditions support Small Water-pepper *Persicaria minor* and the Tasteless Water-pepper *Persicaria mitis*, native members of the knotweed family that have become nationally scarce with the decline in naturally fertile open habitats. The survey of 2008 failed to relocate a Tasteless Water-pepper population discovered less than a decade before; its mudbank habitat had become too saline and whilst apparently amenable berms were present upstream they had not been colonised, at least not yet.

Over the course of a series of surveys between 1996 and 2008 saltmarsh communities were mapped and measured in their migration upstream, the most dynamic achieving a rate of 1.4m per tidal month. The colonisation of the valley by saltmarsh is uneven; watching the tide roll over the grasslands is a daily demonstration of how watercourses accelerate penetration and dense vegetation resists it. Saltmarsh does not advance upstream on an even front; the communities shift across the landscape like bubbles in a lava lamp.

The pace of change at the Lower Test is such that fully freshwater habitats at the head of the marshes are now becoming swampy. There is a zone just above the current reach of the tide supporting Green-winged Orchid *Anacamptis morio* meadows with Quaking-grass

Briza media, Cowslip *Primula veris* and Early Marsh-orchids *Dactylorhiza incarnata* ssp *incarnata*. The challenge in managing this nature reserve is to find somewhere for everything to move to. Fortunately, the valley upstream remains substantially agricultural and is part of the same landed estate as the nature reserve. The Environment Agency has recently entered into an agreement with the estate and the Wildlife Trust to secure future management of the land that will soon be influenced by high tides. Opportunities to meet statutory obligations to make good the adverse effects of sea defences around the Solent are few and far between. Through working with others the Agency has secured such an opportunity, so enabling it to progress its engineered defences along neighbouring urban shores.

The pace of change in the Lower Test has parallels with historical periods of unconstrained saltmarsh migration as recorded by archaeologists in the Humber. Elsewhere in the Solent the seaward face of mature saltmarshes is in retreat; to the west of Southampton Water the saltmarshes of the Lymington River estuary are receding by up to 10m a year, their landward migration prevented by seawalls and natural topography. It has been estimated that there will be no significant saltmarshes on the open shores of the Western Solent by 2050. Current best estimates are that by the end of the century the Hampshire coast will have lost over 800ha of saltmarsh. What is refreshing about the Lower Test marshes is that they are not squeezed; rather, they migrate.

The Lower Test marshes are an exemplum of what can happen if provision is made to accommodate natural dynamism. Saltmarshes readily move upstream if the land is available and if ecological niches are continuously created into which species may colonise.

Rejuvenation

Sooner or later a walk along any coastline where saltmarshes have been embanked will bring you to a seawall that has failed. That failure may be due to an unplanned event or through deliberate decisions; either way the land once enclosed is now exposed to tidal inundation with the reactivation of the process of saltmarsh formation. Seawalls arrest saltmarsh development, a detention that lasts only as long as the wall holds.

The language surrounding changes to a saltmarsh coast is rich in euphemisms reflecting the mixed emotions arising from change. One person's coastal realignment is another's setback; restoration of saltmarsh habitats may be achieved by abandonment of farmland, by surrendering land to the sea. The prefix 'managed' as used in managed retreat and managed realignment brings with it inferences of professional deliberation and control. It is over 25 years since the National Trust decided to lower seawalls on Northey Island in Essex to create saltmarsh from a conventionally farmed grassland. Northey is widely regarded as the first intentional project of its kind by a British conservation body and proved to be the first of many in the Essex marshes. In the early 1990s managed retreat was an acceptable phrase to describe the Northey experiment; today language is more focused on what is to be won through the changes. Whatever the language or motivation, any action or inaction resulting in land being exposed to the tide offers opportunities for the rejuvenation of saltmarshes.

OPPOSITE PAGE:
Spoonbills are emblematic
of restored coastal
wetlands.

Unplanned events

Britain's estuaries abound in the remains of sea defences overwhelmed by circumstance. Most former walls have disappeared altogether, others leave shadowy footings, a line of rubble or a row of stakes pickled in brine; a few are detectable only as administrative ghosts with Ordnance Survey maps exhibiting lost alignments through symbolic rights of way and parish boundaries. Unplanned failures of seawalls still occur, mostly on private land where individuals have to decide whether to invest their own funds in the maintenance of sea defences. The decision to maintain a wall is not solely a matter of economics; land won from the sea by an ancestor is not easily surrendered. There are opportunities in such changes; in the South Hams of Devon an unplanned breach on a coastal estate unexpectedly diversified a tenant's enterprise. Land used for dairying changed to yield a premium crop of *Salicornia* spp. to be sold as organic samphire through a chain of farm shops. Over time the sediments beneath that samphire have consolidated to a point that it is now worth allowing livestock back onto the marsh; business opportunities evolve with the land. The circumstances in this case were exceptional; the financial rewards to those whose land is taken by the sea are rarely so tangible.

On the Gower in Glamorganshire saltmarshes extend along the northern shore of the Burry Inlet. For much of the coast between Llanrhidian and Whiteford the landward limits of the marsh have been embanked, the taller pastures of these grazing marshes contrasting with close-grazed foreshore commons. At the western end of the marshes is Cwm Ivy, an arm of the intertidal embanked and enclosed from the common some 400 years ago.

Cwm Ivy is a property of the National Trust who recognised its potential for restoration to the intertidal as part of their 'Shifting Shores' strategy. In 2013, during the course of discussions into issues and options, the historic sea defences started showing signs of distress. The most vulnerable section of the defences was where the wall crossed an ancient creek, estranged from tidal flows for centuries. In the following August there was catastrophic failure of the seawall; by that winter the creek had reassumed its historical dimensions, so returning nearly 40ha of farmland to the intertidal. The National Trust has chosen not to restore the wall. How the marsh will develop depends on future management; to graze or not to graze, and whether or not to intervene in the evolution of the breach. Livestock retain

OPPOSITE PAGE TOP:
Embanked marshes at Landimore, Llanrhidian marsh.

OPPOSITE PAGE BOTTOM:
The breach at Cwm Ivy, 2016.

access to its western shores where intertidal fens and freshwater flushes are brightened with Yellow Iris *Iris pseudacorus* and marsh-orchids *Dactylorhiza* spp. These communities are reminiscent of those of the western Highlands with the difference that they grade into muddy margins of Annual Sea-blite *Suaeda maritima* and glassworts *Salicornia* spp. with the high probability of colonisation by nearby stands of Mediterranean Sharp Rush *Juncus acutus*.

Measuring success

There are numerous schemes around the British coast where sea defences are being modified to meet legal obligations. In simple terms, if a developer wishes to undertake work that will destroy a saltmarsh of international importance then they must provide an equivalent habitat of equal value that is certain to offset the loss. Similarly, if a government agency wishes to maintain coast defences that will lead to important intertidal habitats being 'squeezed', they too must make good that loss.

An industry has grown up to design and regulate the realignment of coast defences. What was originally a relatively simple inexpensive option has increased in complexity and cost; such changes are inevitable as schemes grow in scale, with inevitable risks to neighbouring properties and commercial interests. A realignment project designed to meet a legal obligation is under particular pressure to demonstrate that, when built, the obligation will be satisfied. Experience gained in the theory and practice of coastal realignments over the last quarter century means that such schemes can be engineered with increasing confidence as to their final physical form. What is still far from predictable is how wildlife takes up the opportunities that the engineers have provided.

Essex Wildlife Trust realigning seawalls on the Colne, 2015.

In a study of nearly 70 sites Hannah Mossman *et al.* (2012) reviewed the character of saltmarshes arising from deliberate coastal engineering together with older unplanned realignments. Their conclusions were that the results were highly variable and only rarely did saltmarshes arising from realignments exhibit the same character and diversity as sites of more natural origin. The saltmarshes instigated through engineering tended to contain a greater than expected abundance of early-stage succession species together with a dominance, once established, of vigorous perennials to the exclusion of more diverse vegetation.

The dominance of a few plants in the developing marsh is unsurprising as there are species that colonise rapidly and occupy the open ground required by other species to become established. Common Reed *Phragmites australis* and Sea-purslane *Atriplex portulacoides* are classic examples of the theoretical concept of competitive exclusion; in the absence of stresses restricting their vigour such coarse perennials will grow to dominate.

In her introduction to site selection Hannah Mossman reported that none of the newly developed saltmarshes were grazed but that some of the sites used for comparison were. On the Sussex coast at Medmerry, a coast defence scheme devised by the Environment Agency and now managed by the RSPB incorporated livestock from the outset. New seawalls have been constructed to defend Selsey that skirt the landward limits of an ancient embanked arm of Pagham Harbour. The new walls were set back from the predicted reach of the intertidal to provide a safety margin within which the revitalised estuary could re-order itself and be colonised by whatever washed in with the tide. Rather than waiting for the inevitable breach of the old defences a mouth was excavated through the shingle banks in September 2013 and the sea was let in.

Sediments undergoing redistribution, Medmerry, 2016.

Species-rich upper saltmarsh, Medmerry, 2016.

After three years the landscape of the revived estuary was still undergoing dramatic changes. The mouth of the harbour, a scoured channel between two shingle spits, migrated landward and determines its own form within the safe embrace of rock armour. Where stock-proof fences have survived this shifting shoreline there are young bullocks that wander across the intertidal with its gentle gradients into the surrounding land. The stocking densities are low and their season restricted; if it were not for their footprints one may not even know they are there. However, the effects of their grazing and trampling are striking; the structure of the vegetation and intervening bare ground is highly diverse. Early-stage succession species are present in abundance, as are potentially dominant species. Amongst suppressed stands of Sea-purslane and Common Cord-grass *Spartina anglica* are Thrift *Armeria maritima*, Sea Rush *Juncus maritimus* and Sea Aster *Aster tripolium*, along with regional rarities including Lax-flowered Sea-lavender *Limonium humile*, Sea Barley *Hordeum marinum* and Sea-heath *Frankenia laevis*. Outside the stock-proof fences Sea Couch *Elytrigia atherica* has already dominated tracts of ungrazed marsh. In moderation the dense grass cover contributes to site diversity and has become the favoured hunting ground of Short-eared Owls *Asio flammeus*.

The engineering of Medmerry accommodates the twice-daily passage of the tides. In contrast, at Alkborough Flats, where the Trent meets the Humber, the engineering has a different objective. Within living memory the Alkborough Flats have been a bombing range and a conventionally managed farm. In the 1950s flood defences were built around 440ha of the marsh to enable its pastures to be converted to arable. In 2006 these walls were adjusted by the Environment Agency to control tidal exchange so that 170ha are permanently inundated, with the remainder of the marsh available to be flooded during extreme events. The capacity of the site to divert flood waters reduces the height of extreme floods in the neighbouring estuary by 15cm, a reduction of risks to local communities valued at £400,000 a year. If the walls had been removed entirely, the flats would be exposed to regular inundation with the consequential accumulation of silt and reduction in capacity. By limiting the number of tides that can access the site the engineers have ensured longevity to their defences.

The volume of freshwater flowing down the Trent means that the estuary at this point is distinctly brackish; Common Reeds have the potential to become dominant in such conditions and Alkborough has developed extensive reedbeds. Within their first decade the reedbeds of the Flats have been adopted by breeding Marsh Harriers *Circus aeruginosus* and Bearded Tits *Panurus biamicus*. The Reed Warbler *Acrocephalus scirpaceus* population can exceed

Short-eared Owls hunting over rank Sea Couch.

300 breeding pairs and the Starling *Sturnus vulgaris* roost has peaked at over 200,000 birds, presenting a spectacular murmuration from the adjacent cliff. The density of reeds reflects the absence of livestock from much of the site; where cattle are able to graze the margins of the marsh they bring with them some structural diversity and species of upper saltmarsh. At the time it was built the Alkborough scheme offered the most extensive opportunity in Britain to rejuvenate the diversity of wetlands latent in an estuary's headwaters. The overall impression to the visitor is of a site managed by and for birdwatchers between the extreme events when it fulfils its primary purpose of safeguarding human life and property.

It was neither legal obligations nor flood protection that inspired the RSPB to undertake Scotland's first coastal realignment in 2003; rather, it was nature conservation. Unusually for saltmarshes north of the Highland line, there is a long history of embanking saltmarshes on the Cromarty Firth. Early encroachments of the intertidal are associated with the monastic settlement at Fearn Abbey. Then, as now, the land created from the estuary was farmed for both livestock and arable, the quality barley contributing to the fame of its local distilleries.

One of the last areas to be embanked from the Cromarty Firth was Meddatt Marsh at Nigg Bay, the wall built in 1950 to create fields of rush pasture. Having taken those fields into their nature reserve, the RSPB re-aligned the wall. There was an almost immediate redistribution of sediments in the bay, with the low ground of the former rush pastures being raised by between 20 and 30cm. Those sediments have been colonised by saltmarsh species, which are settling down to form a grassy saltmarsh of Red Fescue *Festuca rubra* and Common Saltmarsh-grass *Puccinellia maritima*. Cattle grazed this marsh through the early years of restoration but the stock-proof fences became problematic as barriers to bird movement. Posts and wire attracted veils of debris, screening off the marsh from the sandflats. Roosting birds are understandably averse to any restriction to their views of approaching predators. The fence was therefore removed and grazing is now provided by concentrations of wildfowl that gather on the young marsh.

The RSPB is particularly interested in prey species for wintering birds, with Elliott (2015) reporting that by 2011 the Baltic Clams *Macoma balthica* that had colonised the intertidal flats had grown to a size to be attractive to flocks of wintering Knot *Calidris canutus*. As well as feeding Knot, the rejuvenated marsh supports 2,000 waterfowl that resort to the marsh as a high-tide roost; the peak count of roosting Curlew *Numenius arquata* is 800 birds. The newly developed saltmarsh is particularly important during extreme events when the highest tides combine with cold, windy weather. Meddatt Bay is only 20ha but this small area of saltmarsh can play a critical role in supporting waterfowl from far across the estuary.

Curlew roost on the Cromarty Firth.

Of all the realignment schemes I've seen, the best are not over-prescriptive in intent. The most effective engineering makes space for natural processes that can neither be fully predicted nor controlled. Saltmarsh habitats and species colonise intertidal sediments without the need for people to supplement natural processes. Where livestock are integral to a restoration then the fullest range of natural ecological processes are secured, leaving opportunities for the widest range of species to establish. Saltmarshes have a future wherever they have such freedom to move.

The largest of all contemporary British coastal restoration schemes incorporates elements of legal obligations and high-specification design as well as room for the unexpected. The train to Rochford from Liverpool Street station takes you from the commercial heart of the capital through the suburbs and into the Thames estuary. Beyond Rochford an archipelago of marsh islands reaches out towards Foulness and the North Sea. Wallasea, itself an amalgam of smaller islets, is precariously connected to mainland Essex by a single road, inundated on the highest tides.

Wallasea shares a common history with many of its neighbouring saltmarsh islands. When its shoreline was first embanked is unrecorded, but those original defences were incrementally upgraded to eventually support conventional arable cropping. Having been deprived of tidal silt for centuries, the farmland lies well below the peak of high tides. Unlike islands closer to the capital it is almost uninhabited; there are just a few homes clustered around the ferry to Burnham-on-Crouch.

Over the last decade Wallasea has offered a model for the future for saltmarshes. In 2006 its farmland was used to create 118ha of mud and marsh by realigning a stretch of seawall some 400m landward. This was the largest engineered realignment of its time and was required following a High Court judgement concluding that the government had failed to deliver its obligations under European law. In the 1980s estuarine habitats had been destroyed at Fagbury Flats (Suffolk) and Lappell Bank (Kent) without the necessary designations being made nor compensatory measures being provided. Defra, the ministry responsible, handled the project in partnership with commercial consultancies, the Environment Agency and voluntary conservation bodies. In engineering terms the project was a success, and monitoring shows that it has achieved its compensatory targets for birds. As the Wallasea scheme was designed to provide 'two for one' for each hectare of habitat lost, it avoided the risk of inadequate delivery.

Conservation engineering on an unprecedented scale, Wallasea.

Having worked with partners on the project, the RSPB subsequently purchased 700ha of the island and set about delivering the largest coastal habitat recreation project in the country.

When I visited Wallasea in the spring of 2016 it was work in progress. The saltmarsh created a decade before was developing well into a rather uninspiring sward of Sea-purslane *Atriplex portulacoides* and Common Cord-grass *Spartina anglica* which is of equivalent character to its older neighbours. Beyond the new marsh a wharf was being decommissioned, having completed the delivery of over 3 million tonnes of material dug from beneath London. In a mutually advantageous scheme Crossrail had found a beneficial use of arisings from tunnelling their new mainline railway under the capital. In turn, the RSPB welcomed the source of material as they needed to raise the surface of the farmland before flooding it. There was a risk that

Crossrail Place, Canary Wharf. The development of Crossrail provided the sediments to rejuvenate Wallasea's wetlands.

without such land-raising any breach, intentional or accidental, would change the local tidal regime with risks to neighbours and surrounding interests. The land had been raised, sluices had been opened and saltwater was returning to Wallasea after centuries of exclusion.

The scale of the site becomes apparent when you realise it is a good day's ramble (15km) to cover all the perimeter paths and cross dykes. There are areas where the marsh must continue to function in a proscribed way to meet the legacy of legal obligations. Such predetermination of how habitats should develop can be stifling to natural processes if pursued in isolation. Fortunately at Wallasea there is a suite of objectives across a large site with different degrees of management intervention.

Where engineering occurs it is precise; modern groundwork is defined by laser levelling rather than rule of thumb. Opportunities for habitat creation are limited by Wallasea being too small and isolated to have a significant freshwater catchment of its own, and this naturally limits the development of brackish and freshwater features. The scale of Wallasea means that there will be remote places free from disturbance; here bespoke habitat management seeks to tempt Spoonbills *Platalea leucorodia* to breed. After 300 years the birds are returning to England's east coast, but current breeding attempts are sporadic and not always successful. Given the expanding colonies in the Netherlands, it is only a matter of time before they become permanently established in Britain.

Wallasea is large and there is no urgency to determine the precise details of what happens everywhere. There is both the scope and the willingness of the RSPB's team to allow unforeseen circumstances and natural processes to guide future decisions. The suite of management techniques being employed ranges from the restoration of fully intertidal habitats through the regulated tidal exchanges of lagoons and brackish grasslands to wholly terrestrial habitats. It is not the intention to manage Wallasea to restore habitats lost to history. What is being created is a contemporary landscape, rich in wildlife and full of possibilities. Opportunities to work on a blank canvas at such a scale are rare. Wallasea is a model, but it is also an exception; most of the saltmarshes we need to consider are already with us.

In our more industrialised estuaries opportunities to rejuvenate saltmarshes are rare as candidate sites tend to have already been earmarked for more profitable ventures. There is intense competition for development land by the coast and those looking to create estuarine habitats have to compete in that market. It is possible that market forces alone will determine the location of future saltmarshes. In such a scenario our coastal habitats will become increasingly polarised where only estuaries with an extensive rural hinterland will be able to maintain their saltmarsh resource. Such fragmentation of habitat runs contrary to the strategic need for a nationally and internationally coherent suite of wetlands.

When the status of saltmarshes is reviewed the tendency is to focus on size, the number of hectares being used as a proxy for the health of the habitat. It is important to monitor area; current indications are that relative sea-level rise is the leading factor in British saltmarshes shrinking by some 100ha a year. If the objective is to sustain the current area of saltmarsh there is a need to have a rolling programme of schemes such as Cwm Ivy, Nigg and Wallasea.

The concept of shifting baselines was developed in the 1990s by marine biologists who applied it to the management of fisheries, but it is equally applicable to saltmarshes. Shifting baselines occur when a natural resource is in decline but each generation of people regards its own experience as the norm. Over time the diminished resource becomes acceptable and we then fail to strive for its recovery because we no longer appreciate its potential. My experience is that saltmarshes are suffering from shifting baselines; to some, the dullness of a Sea-purslane monoculture or a thatch of marsh grasses is somehow sufficiently good and to be expected. Exemplary saltmarshes may yet inspire us to aspire for excellence.

Time and tide

I n this book I have sought to explore the riches that saltmarshes bring us. As a naturalist I rejoice in that diversity of life; as a conservationist I look to the future.

If we want to safeguard the natural wealth of coastal saltmarshes there is an urgent need to make space for them to move. In the south of Britain this means we must we look to tidal rivers and land behind seawalls to accommodate the migration of intertidal habitats. From a northern perspective the urgency is less intense, but change is coming so it is timely to make ready. Britain's inland saltmarshes are so scarce they are rarely regarded; I hope to have contributed to returning them to their rightful place. I find it extraordinary they have survived at all and salute those who knew and cared about them long before they captured my imagination.

In exploring Britain's saltmarshes I have been concerned about matters of quality. I use quality as shorthand to describe saltmarshes that are complete in their transitions to other habitats and are diverse in the species they support. Most of the places I have described have quality and I hope I have not been too unkind to those that do not. The glory of the inherent dynamism of saltmarshes is that you don't need to wait too long to judge the effects of remedial measures.

With hindsight we conservationists have concentrated on safeguarding large areas of saltmarsh at the expense of considering how best to care for them. Large herbivores are becoming absent from many saltmarshes for the first time in the habitat's evolutionary history. The inevitable consequence of this change is the simplification of ecosystems with the progressive loss of species dependent on

OPPOSITE PAGE:
Arnold's Marsh, Norfolk.

structural diversity. We can try and conserve chosen species through programmes of intensive care, but ultimately this is unsustainable if the habitats they depend upon have been lost. For wildlife to be wild it needs to be able to get along without our constant attention.

For most of British history it is pastoral farming that has provided the human expertise and breeds of livestock that are the ecological successors to our extinct mammal fauna. I would like to think that by redirecting resources we can properly reward those who follow the pastoral traditions of our countryside. There are those who promote a purist approach to rebuilding much wilder landscapes in Britain; this is an interesting notion but it struggles to accommodate people, who have been as much a part of our recent ecological history as any wild species. If re-wilding Britain includes the restoration of a complete native mammal fauna then it has the potential to deliver what pastoral farming is currently doing. I would find proponents of re-wilding more persuasive if they recognised the ways in which extensive agricultural systems have, and do, support exceptional wildlife.

All habitats and landscapes are subject to change, saltmarshes particularly so. At present, the pace of change is overtaking administrative processes and our ability to make decisions. Whilst we discuss our options each tide cycle is just a little higher, there is no status quo to return to nor is inaction free of consequences. In describing change I have striven not to assign human attributes to saltmarshes as in changing they are neither relentless nor indifferent, they have no emotions or intent. Saltmarshes are just a biological manifestation of celestial forces; they are the habitats of time and tide.

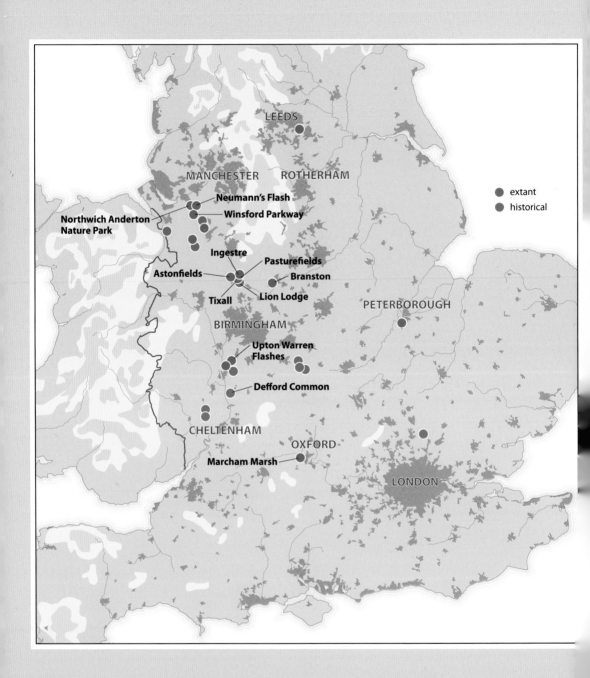

LEEDS

MANCHESTER   ROTHERHAM

Neumann's Flash

Winsford Parkway

Northwich Anderton Nature Park

Ingestre   **Pasturefields**

Astonfields   **Branston**

Tixall   **Lion Lodge**

PETERBOROUGH

BIRMINGHAM

Upton Warren Flashes

Defford Common

CHELTENHAM

OXFORD

Marcham Marsh

LONDON

● extant
● historical

A provisional inventory of inland saltmarshes in Britain

The inland saltmarshes included in this list are wetlands supporting a halophyte flora with a salt source from their underlying geology. Only those sites with two or more such species are listed. Extant species (post 2000 records) are in **bold text**.

The sites are ordered by Watsonian vice-county and then alphabetically by site.

Grid references are as site specific as possible. Where an historical location is unclear the grid reference relates to the nearest settlement.

OPPOSITE PAGE:
The distribution of inland saltmarshes and brackish waters including selected locations.

Site	Flora and fauna	Geology	Data source
Hertfordshire			
Ware Park Brickfields TL 34 15	Lesser Sea-spurrey *Spergularia marina* Reflexed Saltmarsh-grass *Puccinellia distans*	London Clay	Dony 1967
Berkshire			
Marcham Marsh SU 453 960	Strawberry Clover *Trifolium fragiferum* Lesser Sea-spurrey *Spergularia marina* Brookweed *Samolus valerandi* Spear-leaved Orache *Atriplex prostrata* Sea Arrowgrass *Triglochin maritima* Parsley Water-dropwort *Oenanthe lachenalii* **Wild Celery** **Apium graveolens** **Saltmarsh Rush** **Juncus gerardii**	Jurassic formations Junction of Stanford Formation limestones and mudstones of Amphill/Kimmeridge/ Hazlebury Bryan formations	Druce 1897 and Oxfordshire Flora Group pers. comm.

Site	Flora and fauna	Geology	Data source
Berkshire cont.			
	Sea Club-rush *Bolboschoenus maritimus* **Distant Sedge** *Carex distans*		
Huntingdonshire			
Orton Pits TL 163 940	**Golden Dock** *Rumex maritimus* **Lesser Pondweed** *Potamogeton pusillus* **Fennel Pondweed** *Potamogeton pectinatus* Charophytes (Algae) *Chara canescens* *Chara aspera* *Chara pedunculata* *Tolypella glomerata*	London Clay	Natural England SSSI citation
Gloucestershire			
Ketford and River Leadon SO 729 306	Sea Beet *Beta vulgaris* ssp *maritima* Sea Arrowgrass *Triglochin maritima* Wild Celery *Apium graveolens* Saltmarsh Rush *Juncus gerardii* Sea Club-rush *Bolboschoenus maritimus* Distant Sedge *Carex distans*	Triassic Mercia Mudstones Group	Riddelsdell *et al.* 1948
Newent SO 728 257	Brookweed *Samolus valerandi* Wild Celery *Apium graveolens*	Triassic Mercia Mudstones Group	Riddelsdell *et al.* 1948
Worcestershire			
Defford Common SO 920 431	Lesser Sea-spurrey *Spergularia marina* Parsley Water-dropwort *Oenanthe lachenalii* Slender Hare's-ear *Bupleurum tenuissimum* Sea Club-rush *Bolboschoenus maritimus*	Triassic Mercia Mudstones Group	Amphlett & Rea 1909
Droitwich Canal (Salwarpe) SO 876 618	**Dittander** *Lepidium latifolium* Lesser Sea-spurrey *Spergularia marina*	Triassic Mercia Mudstones Group	Amphlett & Rea 1909 Worcestershire Biological Records Centre CC

Site	Flora and fauna	Geology	Data source
	Sea-milkwort *Glaux maritima* Wild Celery *Apium graveolens* Saltmarsh Rush *Juncus gerardii* Distant Sedge *Carex distans* Common Saltmarsh-grass *Puccinellia maritima* **Reflexed Saltmarsh-grass** ***Puccinellia distans*** Sea Barley *Hordeum marinum*		
Saldon near Himbledon SO 940 584	Babington's Orache *Atriplex glabriuscula* Saltmarsh Rush *Juncus gerardii*	Triassic Mercia Mudstones Group	Amphlett & Rea 1909
Upton Warren Flashes SO 934 666	**Spear-leaved Orache** ***Atriplex prostrata*** **Lesser Sea Spurrey** ***Spergularia marina*** **Grey Club-rush** ***Schoenoplectus tabernaemontani*** **Reflexed Saltmarsh-grass** ***Puccinellia distans*** **Heim's Pottia** (a moss) ***Hennediella heimii*** **Avocet** (breeding) ***Recurvirostra avosetta*** Coleoptera ***Enochrus bicolour*** ***Anthicus constrictus*** ***Bembidion varium*** ***Saldula pilosella*** Diptera ***Melanum laterale*** ***Molophilus pleuralis*** ***Stratiomys singularior***	Triassic Mercia Mudstones Group	Amphlett & Rea 1909 Worcestershire Wildlife Trust & Biological Records Centre CC Coleoptera and Diptera records Don Goddard & David M Green
Warwickshire			
Itchington Holt (c. SP 41 65)	Sea Club-rush *Bolboschoenus maritimus* Grey Club-rush *Schoenoplectus tabernaemontani*	Triassic Mercia Mudstones Group	Cadbury *et al.* 1971

Site	Flora and fauna	Geology	Data source
Warwickshire cont.			
Napton SP 444 602	Sea Club-rush *Bolboschoenus maritimus* Grey Club-rush *Schoenoplectus tabernaemontani*	Triassic Mercia Mudstones Group	Lee 1977
Southam Holt (c. SP 42 61)	Wild Celery *Apium graveolens* Sea Club-rush *Bolboschoenus maritimus* Grey Club-rush *Schoenoplectus tabernaemontani*	Triassic Mercia Mudstones Group	Cadbury *et al.* 1971
Staffordshire			
Astonfields SJ 926 248	**Spear-leaved Orache** ***Atriplex prostrata*** **Lesser Sea Spurrey** ***Spergularia marina*** **Buttonweed** ***Cotula coronopifolia*** (non-native) **Grey Club-rush** ***Schoenoplectus tabernaemontani*** Reflexed Saltmarsh-grass *Puccinellia distans*	Triassic Mercia Mudstones Group	Hawksford & Hopkins 2011 Staffordshire Wildlife Trust pers. comm. Bill Waller pers. comm. CC
Branston SK 222 206	Brackish Water-crowfoot *Ranunculus baudotii* **Golden Dock** ***Rumex maritimus*** Lesser Sea-spurrey *Spergularia marina* Spear-leaved Orache *Atriplex prostrata* Sea-milkwort *Glaux maritima* **Brookweed** ***Samolus valerandi*** Sea Aster *Aster tripolium* Wild Celery *Apium graveolens* Sea Arrowgrass *Triglochin maritima* Grass-wrack Pondweed *Potamogeton compressus* **Saltmarsh Rush** ***Juncus gerardii*** Sea Club-rush *Bolboschoenus maritimus*	Triassic Mercia Mudstones Group	Lee 1977 Edees 1972 Hawksford & Hopkins 2011 Mike Smith pers. comm.

Site	Flora and fauna	Geology	Data source
	Grey Club-rush *Schoenoplectus tabernaemontani* Common Saltmarsh-grass *Puccinellia maritima*		
Ingestre SJ 980 247	Sea Aster *Aster tripolium* Sea Arrowgrass *Triglochin maritima*	Triassic Mercia Mudstones Group	Edees 1972 Hawksford & Hopkins 2011
Kingston Pool (Stafford) SJ 944 235	Lesser Sea-spurrey *Spergularia marina* Saltmarsh Rush *Juncus gerardii* Sea Club-rush *Bolboschoenus maritimus*	Triassic Mercia Mudstones Group	Edees 1972 Hawksford & Hopkins 2011
Lion Lodge SJ 989 239	**Saltmarsh Rush** *Juncus gerardii* **Borrer's Saltmarsh-grass** *Puccinellia fasciculata*	Triassic Mercia Mudstones Group	Dave Cadnam pers. comm.
Pasturefields SJ 992 248	**Lesser Sea-spurrey** *Spergularia marina* **Spear-leaved Orache** *Atriplex prostrata* **Sea-milkwort** *Glaux maritima* **Sea Plantain** *Plantago maritima* **Sea Arrowgrass** *Triglochin maritima* **Saltmarsh Rush** *Juncus gerardii* Common Saltmarsh-grass *Puccinellia maritima*	Triassic Mercia Mudstones Group	Hawksford & Hopkins 2011 Staffordshire Wildlife Trust pers. comm. Sue Lawley and CC
Shirleywich SJ 984 259	Wild Celery *Apium graveolens* **Sea Arrowgrass** *Triglochin maritima* Saltmarsh Rush *Juncus gerardii*	Triassic Mercia Mudstones Group	Edees 1972 Hawksford & Hopkins 2011 Staffordshire Wildlife Trust pers. comm.
Tixall SJ 976 227	Lesser Sea-spurrey *Spergularia marina* Brookweed *Samolus valerandi* Sea-milkwort *Glaux maritima* Sea Plantain *Plantago maritima* Sea Aster *Aster tripolium* Sea Arrowgrass *Triglochin maritima*	Triassic Mercia Mudstones Group	Lee 1977 Hawksford & Hopkins 2011 Clifford & Clifford 1817

Site	Flora and fauna	Geology	Data source
Cheshire			
Aldersey SJ 457 567	Spear-leaved Orache *Atriplex prostrata* Sea Arrowgrass *Triglochin maritima* Saltmarsh Rush *Juncus gerardii*	Triassic Mercia Mudstones Group	Lee 1977
Audlem SJ 664 466	Spear-leaved Orache *Atriplex prostrata* Lesser Sea-spurrey *Spergularia marina* Reflexed Saltmarsh-grass *Puccinellia distans*	Triassic Mercia Mudstones Group	Lee 1977
Winsford Parkway SJ 655 681	**Lesser Sea-spurrey** ***Spergularia marina*** **Spear-leaved Orache** ***Atriplex prostrata*** **Sea-milkwort** ***Glaux maritima*** **Brookweed** ***Samolus valerandi*** **Sea Aster** ***Aster tripolium*** **Grey Club-rush** ***Schoenoplectus tabernaemontani*** **Reflexed Saltmarsh-grass** ***Puccinellia distans***	Triassic Mercia Mudstones Group	Lee 1977 Graeme Kay pers. comm. CC
Nantwich Brine Springs Hotel SJ 651 509	Lesser Sea-spurrey *Spergularia marina* Spear-leaved Orache *Atriplex prostrata* Babington's Orache *Atriplex glabriuscula* Sea Arrowgrass *Triglochin maritima* Saltmarsh Rush *Juncus gerardii* Common Saltmarsh-grass *Puccinellia maritima* Reflexed Saltmarsh-grass *Puccinellia distans*	Triassic Mercia Mudstones Group	Lee 1975 and 1977 Graeme Kay pers. comm.
Nantwich Lake SJ 648 514	**Lesser Sea-spurrey** ***Spergularia marina*** **Spear-leaved Orache** ***Atriplex prostrata*** Sea Arrowgrass *Triglochin maritima* Beaked Tasselweed *Ruppia maritima*	Triassic Mercia Mudstones Group	Lee 1975 and 1977 Graeme Kay pers. comm. CC

Site	Flora and fauna	Geology	Data source
	Saltmarsh Rush *Juncus gerardii* **Reflexed Saltmarsh-grass** *Puccinellia distans*		
Nantwich Fields Farm SJ 648 514	Lesser Sea-spurrey *Spergularia marina* Spear-leaved Orache *Atriplex prostrata* Babington's Orache *Atriplex glabriuscula* Sea-milkwort *Glaux maritima* Brookweed *Samolus valerandi* Sea Arrowgrass *Triglochin maritima* Beaked Tasselweed *Ruppia maritima* Saltmarsh Rush *Juncus gerardii* Grey Club-rush *Schoenoplectus tabernaemontani* Common Saltmarsh-grass *Puccinellia maritima* Reflexed Saltmarsh-grass *Puccinellia distans*	Triassic Mercia Mudstones Group	Lee 1975 and 1977
Northwich SJ 66 74	Lesser Sea-spurrey *Spergularia marina* Spear-leaved Orache *Atriplex prostrata* Reflexed Saltmarsh-grass *Puccinellia distans*	Triassic Mercia Mudstones Group	Lee 1977
Northwich Anderton Nature Park SJ 651 751	**Lesser Sea-spurrey** *Spergularia marina* **Spear-leaved Orache** *Atriplex prostrata* **Wild Celery** *Apium graveolens* **Stiff Saltmarsh-grass** *Puccinellia rupestris* **Reflexed Saltmarsh-grass** *Puccinellia distans*	Triassic Mercia Mudstones Group	Lee 1977 Graeme Kay pers. comm.
Neumann's Flash SJ 667 750	Purple Glasswort *Salicornia ramosissima* **Sea Aster** *Aster tripolium* **Horned Pondweed** *Zannichellia palustris* **Reflexed Saltmarsh-grass** *Puccinellia distans*	Triassic Mercia Mudstones Group	Graeme Kay pers. comm. CC

Site	Flora and fauna	Geology	Data source
Cheshire cont.			
Sandbach Flashes SJ 725 594	**Lesser Sea-spurrey** *Spergularia marina* **Spear-leaved Orache** *Atriplex prostrata* **Sea Aster** *Aster tripolium* Beaked Tasselweed *Ruppia maritima* **Grey Club-rush** *Schoenoplectus tabernaemontani* **Reflexed Saltmarsh-grass** *Puccinellia distans* *Enteromorpha intestinalis* (an alga) Hemiptera *Sigara conicinna* *Sigara stagnalis* Crustacea *Gammarus duebeni* *Gammarus tririnus*	Triassic Mercia Mudstones Group	Lee 1977 Natural England SSSI citation
Silver Spring (Anderton) SJ 652 750	Spear-leaved Orache *Atriplex prostrata* Sea Club-rush *Bolboschoenus maritimus* Reflexed Saltmarsh-grass *Puccinellia distans*	Triassic Mercia Mudstones Group	Lee 1977
Cledford (Middlewich) SJ 710 647	Lesser Sea-spurrey *Spergularia marina* Reflexed Saltmarsh-grass *Puccinellia distans*	Triassic Mercia Mudstones Group	Newton 1971
Yorkshire			
Mickletown Ings SE 403 275	**Grey Club-rush** *Schoenoplectus tabernaemontani* **Saltmarsh Rush** *Juncus gerardii* Crustacea *Gammarus duebeni* Coleoptera *Macroplea mutica* Diptera *Porphyrops antennae*	Carboniferous coal measures	Natural England SSSI citation

CC: Personal visit by the author

Special Areas of Conservation in Britain supporting saltmarsh habitats

SAC name	Local Authority	Saltmarsh features present within the SAC
Afon Teifi/River Teifi	Caerfyrddin/Carmarthenshire; Ceredigion; Penfro/Pembrokeshire	1130 Estuaries 1140 Mudflats and sandflats not covered by seawater at low tide 1330 Atlantic salt meadows
Alde, Ore and Butley Estuaries	Suffolk	1130 Estuaries 1140 Mudflats and sandflats not covered by seawater at low tide 1330 Atlantic salt meadows
Ardvar and Loch a' Mhuilinn Woodlands	Highland	1330 Atlantic salt meadows
Bae Cemlyn/Cemlyn Bay	Ynys Môn/Isle of Anglesey	1150 Coastal lagoons
Benacre to Easton Bavents Lagoons	Suffolk	1150 Coastal lagoons
Berwickshire and North Northumberland Coast	Northumberland; Scottish Borders	1130 Estuaries 1140 Mudflats and sandflats not covered by seawater at low tide 1150 Coastal lagoons 1310 *Salicornia* and other annuals colonising mud and sand 1320 *Spartina* swards 1330 Atlantic salt meadows
Buchan Ness to Collieston	Aberdeenshire	1230 Vegetated sea cliffs of the Atlantic and Baltic coasts
Cape Wrath	Highland	1230 Vegetated sea cliffs of the Atlantic and Baltic coasts

SAC name	Local Authority	Saltmarsh features present within the SAC
Carmarthen Bay and Estuaries/ Bae Caerfyrddin ac Aberoedd	Abertawe/Swansea; Caerfyrddin/Carmarthenshire; Penfro/Pembrokeshire	1130 Estuaries 1140 Mudflats and sandflats not covered by seawater at low tide 1310 *Salicornia* and other annuals colonising mud and sand 1330 Atlantic salt meadows
Chesil and the Fleet	Dorset	1150 Coastal lagoons 1330 Atlantic salt meadows 1420 Mediterranean and thermo-Atlantic halophilous scrubs
Claish Moss and Kentra Moss	Highland	1130 Estuaries 1140 Mudflats and sandflats not covered by seawater at low tide 1330 Atlantic salt meadows
Conon Islands	Highland	91E0 Alluvial forests with *Alnus glutinosa* and *Fraxinus excelsior*
Culbin Bar	Highland; Moray	1330 Atlantic salt meadows
Dee Estuary/ Aber Dyfrdwy	Cheshire; Sir y Fflint/Flintshire; Wirral	1130 Estuaries 1140 Mudflats and sandflats not covered by seawater at low tide 1310 *Salicornia* and other annuals colonising mud and sand 1330 Atlantic salt meadows
Dornoch Firth and Morrich More	Highland	1110 Sandbanks which are slightly covered by seawater all the time 1130 Estuaries 1140 Mudflats and sandflats not covered by seawater at low tide 1310 *Salicornia* and other annuals colonising mud and sand 1330 Atlantic salt meadows
Drigg Coast	Cumbria	1130 Estuaries 1140 Mudflats and sandflats not covered by seawater at low tide 1310 *Salicornia* and other annuals colonising mud and sand 1330 Atlantic salt meadows
East Caithness Cliffs	Highland	1230 Vegetated sea cliffs of the Atlantic and Baltic coasts
Essex Estuaries	Essex	1110 Sandbanks which are slightly covered by sea water all the time 1130 Estuaries 1140 Mudflats and sandflats not covered by seawater at low tide 1310 *Salicornia* and other annuals colonising mud and sand

SAC name	Local Authority	Saltmarsh features present within the SAC
Essex Estuaries cont.	Essex	1320 *Spartina* swards 1330 Atlantic salt meadows 1420 Mediterranean and thermo-Atlantic halophilous scrubs

Fal and Helford	Cornwall	1110 Sandbanks which are slightly covered by
		    seawater all the time
		1130 Estuaries
		1140 Mudflats and sandflats not covered by
		    seawater at low tide
		1330 Atlantic salt meadows

Firth of Tay and Eden Estuary	Angus;
	City of Dundee;
	Fife;
	Perth & Kinross
		1110 Sandbanks which are slightly covered by
		    seawater all the time
		1130 Estuaries
		1140 Mudflats and sandflats not covered by
		    seawater at low tide

Glannau Môn: Cors heli/ Anglesey Coast: Saltmarsh	Ynys Môn/Isle of Anglesey	1130 Estuaries
		1140 Mudflats and sandflats not covered by
		    seawater at low tide
		1310 *Salicornia* and other annuals colonising
		    mud and sand
		1330 Atlantic salt meadows

Hamford Water	Essex	1130 Estuaries
		1140 Mudflats and sandflats not covered by
		    seawater at low tide
		1310 *Salicornia* and other annuals colonising
		    mud and sand
		1320 *Spartina* swards
		1330 Atlantic salt meadows

Humber Estuary	City of Kingston upon Hull;
	East Riding of Yorkshire;
	Lincolnshire;
	North East Lincolnshire;
	North Lincolnshire
		1110 Sandbanks which are slightly covered by
		    seawater all the time
		1130 Estuaries
		1140 Mudflats and sandflats not covered by
		    seawater at low tide
		1150 Coastal lagoons
		1310 *Salicornia* and other annuals colonising
		    mud and sand
		1330 Atlantic salt meadows

Invernaver	Highland	1130 Estuaries 1140 Mudflats and sandflats not covered by seawater at low tide
Isle of Wight Downs	Isle of Wight	1230 Vegetated sea cliffs of the Atlantic and Baltic coasts
Kenfig/Cynffig	Bro Morgannwg/Vale of Glamorgan; Pen-y-bont ar Ogwr/Bridgend	1330 Atlantic salt meadows
Loch Laxford	Highland	1330 Atlantic salt meadows

SAC name	Local Authority	Saltmarsh features present within the SAC
Loch Moidart and Loch Shiel Woods	Highland	1140 Mudflats and sandflats not covered by seawater at low tide 1330 Atlantic salt meadows
Loch nam Madadh	Western Isles/Na h-Eileanan Siar	1110 Sandbanks which are slightly covered by seawater all the time 1140 Mudflats and sandflats not covered by seawater at low tide 1150 Coastal lagoons
Loch of Stenness	Orkney Islands	1150 Coastal lagoons
Loch Roag Lagoons	Western Isles/Na h-Eileanan Siar	1150 Coastal lagoons
Mòine Mhòr	Argyll and Bute	1140 Mudflats and sandflats not covered by seawater at low tide 1330 Atlantic salt meadows
Morecambe Bay	Cumbria; Lancashire	1110 Sandbanks which are slightly covered by seawater all the time 1130 Estuaries 1140 Mudflats and sandflats not covered by seawater at low tide 1150 Coastal lagoons 1310 *Salicornia* and other annuals colonising mud and sand 1330 Atlantic salt meadows
Morfa Harlech a Morfa Dyffryn	Gwynedd	1140 Mudflats and sandflats not covered by seawater at low tide 1310 *Salicornia* and other annuals colonising mud and sand 1330 Atlantic salt meadows
Mound Alderwoods	Highland	91E0 Alluvial forests with *Alnus glutinosa* and *Fraxinus excelsior*
North Norfolk Coast	Norfolk	1140 Mudflats and sandflats not covered by seawater at low tide 1150 Coastal lagoons 1310 *Salicornia* and other annuals colonising mud and sand 1330 Atlantic salt meadows 1420 Mediterranean and thermo-Atlantic halophilous scrubs
North Uist Machair	Western Isles/Na h-Eileanan Siar	1330 Atlantic salt meadows
Obain Loch Euphoirt	Western Isles/Na h-Eileanan Siar	1150 Coastal lagoons
Orfordness–Shingle Street	Suffolk	1150 Coastal lagoons

SAC name	Local Authority	Saltmarsh features present within the SAC
Orton Pit	City of Peterborough	3140 Hard oligo-mesotrophic waters with benthic vegetation of *Chara* spp.
Pasturefields Salt Marsh	Staffordshire	1340 Inland salt meadows
Pembrokeshire Marine/ Sir Benfro Forol	Penfro/Pembrokeshire	1110 Sandbanks which are slightly covered by seawater all the time 1130 Estuaries 1140 Mudflats and sandflats not covered by seawater at low tide 1150 Coastal lagoons 1330 Atlantic salt meadows
Pen Llyn a`r Sarnau/ Lleyn Peninsula and the Sarnau	Ceredigion; Gwynedd; Powys	1130 Estuaries 1140 Mudflats and sandflats not covered by seawater at low tide 1150 Coastal lagoons 1310 *Salicornia* and other annuals colonising mud and sand 1330 Atlantic salt meadows
Plymouth Sound and Estuaries	Cornwall; Devon; Plymouth	1110 Sandbanks which are slightly covered by seawater all the time 1130 Estuaries 1140 Mudflats and sandflats not covered by seawater at low tide 1330 Atlantic salt meadows
Sandwich Bay	Kent	1130 Estuaries 1140 Mudflats and sandflats not covered by seawater at low tide 1330 Atlantic salt meadows
Severn Estuary/ Môr Hafren	Bro Morgannwg/Vale of Glamorgan; Caerdydd/Cardiff; Casnewydd/Newport; City of Bristol; Fynwy/ Monmouthshire; Gloucestershire; North Somerset; Somerset; South Gloucestershire	1110 Sandbanks which are slightly covered by seawater all the time 1130 Estuaries 1140 Mudflats and sandflats not covered by seawater at low tide 1330 Atlantic salt meadows
Sidmouth to West Bay	Devon; Dorset	1230 Vegetated sea cliffs of the Atlantic and Baltic coasts
Solent and Isle of Wight Lagoons	City of Portsmouth; Hampshire; Isle of Wight	1150 Coastal lagoons
Solent Maritime	City of Portsmouth; City of Southampton; Hampshire; Isle of Wight; West Sussex	1110 Sandbanks which are slightly covered by seawater all the time 1130 Estuaries 1140 Mudflats and sandflats not covered by seawater at low tide 1150 Coastal lagoons 1310 *Salicornia* and other annuals colonising mud and sand 1320 *Spartina* swards 1330 Atlantic salt meadows

SAC name	Local Authority	Saltmarsh features present within the SAC
Solway Firth	Cumbria; Dumfries and Galloway	1110 Sandbanks which are slightly covered by seawater all the time 1130 Estuaries 1140 Mudflats and sandflats not covered by seawater at low tide 1310 *Salicornia* and other annuals colonising mud and sand 1330 Atlantic salt meadows
Sound of Arisaig (Loch Ailort to Loch Ceann Traigh)	Highland	1110 Sandbanks which are slightly covered by seawater all the time
South Devon Shore Dock	Devon	1230 Vegetated sea cliffs of the Atlantic and Baltic coasts
South Uist Machair	Western Isles/Na h-Eileanan Siar	1150 Coastal lagoons
Sullom Voe	Shetland Islands	1150 Coastal lagoons
Taynish and Knapdale Woods	Argyll and Bute	1330 Atlantic salt meadows
Tayvallich Juniper and Coast	Argyll and Bute	1330 Atlantic salt meadows
The Broads	Norfolk; Suffolk	3140 Hard oligo-mesotrophic waters with benthic vegetation of *Chara* spp.
The Vadills	Shetland Islands	1150 Coastal lagoons
The Wash and North Norfolk Coast	Lincolnshire; Norfolk	1110 Sandbanks which are slightly covered by seawater all the time 1140 Mudflats and sandflats not covered by seawater at low tide 1150 Coastal lagoons 1310 *Salicornia* and other annuals colonising mud and sand 1330 Atlantic salt meadows 1420 Mediterranean and thermo-Atlantic halophilous scrubs
Tweed Estuary	Northumberland	1130 Estuaries

The numbering of the features reflects the Natura 2000 standard vocabulary.

The saltmarsh features listed as present within each location may not be the reason for its selection as a SAC.

1110 and 1140 features are included where *Zostera* or *Ruppia* beds form part of a larger saltmarsh complex.

1230 features are included where perched saltmarshes are present. Data are inadequate to determine whether this list is complete.

Many sites have additional non-saltmarsh SAC features that are not listed here.

References and further reading

Adam, P 1990 *Saltmarsh ecology.* Cambridge University Press, Cambridge

Adnitt, C, Brew, D, Cottle, R, Hardwick, M, John, S, Leggett, D, McNulty, S, Meakins, N, & Staniland, R 2007 *Saltmarsh management manual.* Science Report – R&D Technical Report (PFA-076/TR). Environment Agency, Bristol

Allen, D E 1996 Some early workers in the Hampshire flora. In: Brewis, A, Bowman, P & Rose, F *The Flora of Hampshire.* Harley Books, Colchester

Allen, D E 1986 *The Botanists.* St Paul's Bibliographies, Winchester

Allen, J R L 1986 A short history of salt-marsh reclamation at Slimbridge Warth and neighbouring areas, Gloucestershire. *Transactions of the Bristol and Gloucestershire Archaeological Society* 104: 139–155

Allen, J R L 2001 Land-claim and sea defence: labour costs of historical earth banks in Holocene coastal lowlands, NW Europe *Archaeology in the Severn Estuary* 12: 127–134

Allen, J R L, & Fulford, M G 1990 Romano-British and later reclamations on the Severn salt marshes in the Elmore area, Gloucestershire. *Transactions of the Bristol and Gloucestershire Archaeological Society* 108: 17–32

Allen, J R L, & Pye, K (eds.) 1992 *Saltmarshes: Morphodynamics, conservation and engineering significance.* Cambridge University Press, Cambridge

Allen, J R L & Rippon, S J 1995 The historical simplification of coastal flood defences: four case histories from the Severn Estuary Levels.

Transactions of the Bristol and Gloucestershire Archaeological Society 113: 73–88

Amphlett, J, & Rea, C 1909 *The botany of Worcestershire.* Cornish Brothers, Birmingham

An, S Q, Gu, B H, Zhou, C F, Wang, Z S, Deng, Z F, Zhi, Y B, Li, H L, Chen, L, Yu, D H & Liu, Y H 2007 *Spartina* invasion in China: implications for invasive species management and future research. *Weed Research* 47: 183–191

Anon 1607 *God's warning to his people in England.* Barley & Bayly, London

Asem, A 2008 Historical record of Brine Shrimp *Artemia* more than one thousand years ago from Urmia Lake, Iran. *Journal of Biological Research – Thessaloniki* 9: 113–114

Babington, C C 1860 *Flora of Cambridgeshire.* John Van Voorst, London

Baker, J G 1906 *North Yorkshire: studies of its botany, geology, climate and physical geography.* Brown and Sons, London

Bakker, D 1960 *Senecio congestus* in the Lake Yssel Polders. *Acta Botanica Neerlandica* 9: 235–259

Balfour-Browne, F 1948 Re-discovery of *Apus cancriformis. Nature* 17 July 1948, 162: 116

Bamber, R N 2010 *Coastal saline lagoons and the Water Framework Directive.* Natural England Commissioned Report NECR039

Baring-Gould, S 1880 *Mehalah, a story of the salt marshes.* Smith Elder & Co, London

Baring-Gould, S 1894 *The Queen of Love.* Methuen, London

Beaumont, N J, Jones, L, Garbutt, A, Hansom, J D & Tobermann, M 2014 The value of carbon sequestration and storage in coastal habitats. *Estuarine, Coastal and Shelf Science* 137: 32–40

Becker, D, & Stills, N 1988 *The conversion of saltmarsh into fresh and brackish water habitats at Titchwell Marsh, Norfolk.* Unpublished Management Case Study. RSPB, Sandy

Beckett, G, & Bull, A 1999 *A Flora of Norfolk.* Jarrold, Norwich

Bennett, M 1991 Plant lore in Gaelic Scotland. In: Pankhurst, R J & Mullin, J M *Flora of the Outer Hebrides.* The Natural History Museum, London

Body, R 1983 *Agriculture, the triumph and the shame.* Temple Smith, London

Boorman, L A 2003 *Saltmarsh review.* JNCC Report No.334. JNCC, Peterborough

Booy, O, Wade M, & Roy, H 2015 *Field guide to invasive plants and animals in Britain.* Bloomsbury, London

Botterna, S, & Cappers, R T J 2003 A reconstruction of the landscape of the mammoth site near Orvelte, the Netherlands. In: Reumer, J W F, de Vos, J & Mol, D (eds.) *Advances in mammoth research. Deinsea* 9: 87–95

Boudot, J-P 2014 *Coenagrion scitulum.* IUCN red list of threatened species. On: www.iucnredlist.org

Broads Authority 2011 *Broads plan.* Broads Authority, Norwich

Bromfield, W A 1836 A description of *Spartina alterniflora* of Loiseleur, A new British species. In: Hooker, W J (ed) *Companion to the botanical magazine.* Samuel Curtis, London

Bromfield, W A 1856 *Flora Vectensis.* William Pamplin, London

BSBI *News* archives at http://archive.bsbi.org.uk/bsbi_news.html

Burd, F 1989 *Saltmarsh survey of Great Britain: Research and survey in nature conservation* No. 17. NCC, Peterborough

Buxton, A 1938 Sea-sodden Horsey. *The Spectator* 15 September 1938

Buxton, Lord 1981 Draining Broadland, a recipe for conflict and taxpayers' involuntary profligacy. Privately printed correspondence with agriculture ministers

Cadbury, D A, Hawkes, J G & Readett, R C C 1971 *A computer-mapped flora. A study of the county of Warwickshire.* BNHS London: Academic Press

Cameron, J 1883 *The Gaelic names of plants.* John Mackay, Glasgow

Carlisle, D B 1968 *Triops* (Entomostraca) eggs killed only by boiling. *Science* 161: 279–280

Cary, E 1914–27 (translation of Cassius Dio) *Historia Romana.* Loeb Classical Library. Harvard University Press, Harvard

Chapman, P 2007 *Conservation grazing of semi-natural habitats:* TC 586. SAC, Edinburgh

Chatters, C 2004 Grazing domestic animals on British saltmarshes. *British Wildlife* 15(6): 392–400

Cheffings, C M, & Farrell, L (eds.) 2006 *The vascular plant red data list for Great Britain.* JNCC, Peterborough

Chevalier, M, Pye, S, Porter, J & Chambers, S 2014 *Hydrobiidae of North Uist.* SNH Commissioned Report No.559

Clifford, T, & Clifford, A 1817 *A topographical and historical description of the Parish of Tixall in the County of Stafford.* M. Nouzou, Paris

Cole, A, Cumber J, & Gelling, M 2000 Old English merece 'Wild Celery, Smallage' in place names. *Nomina* 23: 141–148

Coles, B J 1998 Doggerland: A speculative survey. *Proceedings of Prehistoric Society* 64: 45–81

Colt Hoare, R 1806 *The itinerary of Archbishop Baldwin through Wales AD 1188 by Giraldus de Barri.* London

Conrad, J 1899 *Heart of Darkness.* Blackwood's Magazine, London

Coope, G R, Shotton, F W, Strachan, I & Dance, S P 1961 A Late Pleistocene Fauna and Flora from Upton Warren, Worcestershire. *Philosophical Transactions of the Royal Society of London, Series B* 244(714): 379–421

Cope, T & Gray, A 2009 *Grasses of the British Isles: BSBI Handbook No. 13.* BSBI, London

Culpeper, N 1653 *The complete herbal.* London

Darby, H C 1940 *The draining of the fens.* Cambridge University Press, Cambridge

Darwin, C 1839 *Voyages of the Adventure and Beagle. Journal and Remarks 1832–1836.* Henry Colburn, London

Davidson, N C, d'A Laffoley, D, Doody, J P, Way, L S, Gordon, J, Key, R, Pienkowski, M W, Mitchell, R & Duff, K L 1991 *Nature conservation and estuaries in Great Britain.* NCC, Peterborough

Dean, M, Ashton, P A, Hutcheon, K, Jermy, A C & Cayouette, J 2008 Description, ecology and establishment of *Carex salina* Wahlenb. (Saltmarsh Sedge) – a new British Species. *Watsonia* 27: 51–57

Denman, D R, Roberts, R A & Smith, H J F 1967 *Commons and village greens*. Leonard Hill, London

Dickens, C 1861 *Great Expectations*. Chapman and Hall, London

Dickens, C 1864 *Our Mutual Friend*. Chapman and Hall, London

Dickson, J H 1992 North American driftwood, especially *Picea* (Spruce) from archaeological sites in the Hebrides and Northern Isle of Scotland. *Review of Paleobotany and Palynology* 73: 49–56

Dijkema, K S (ed.) 1984 *Salt marshes in Europe*. Nature and Environment series, No.30. Council of Europe, Strasbourg

Dony, J G 1967 *Flora of Hertfordshire*. Hitchin Museum

Doody, J P 2001 *Coastal conservation and management: an ecological perspective*. Kluwer, Academic Publishers, Boston

Doody, J P 2011 *Saltmarsh conservation, management and restoration*. Springer-Verlag, New York

Doody, P, & Barnett, B (eds.) 1987 *The Wash and its environment: Research and Survey in Nature Conservation no.7*. NCC. Peterborough

Drayton, M 1622 *Poly-olbion*. London

Druce, G C 1932 *The comital flora of the British Isles*. Buncle & Co, Arbroath

Druce, G C 1897 *The Flora of Berkshire*. Clarendon Press, Oxford

Duigan, C A, Kovach, W L & Palmer, M 2006 *Vegetation communities of British lakes; a revised classification*. JNCC, Peterborough

Edees, E S 1972 *Flora of Staffordshire*. David and Charles, Newton Abbot

Elliott, S 2015 *Coastal realignment at RSPB Nigg Bay nature reserve*. RSPB Internal Report

Environment Agency 2011 *Severn estuary flood risk management strategy: Strategic environmental assessment report*

Evans, P A, Evans, I M & Rothero, G P 2002 *Flora of Assynt*. Evans and Evans, Belfast

Foster, G 2000 The aquatic Coleoptera of British saltmarshes. In: Sherwood, B R, Gardener B G, & Harris, T (eds.) *British saltmarshes*. Linnean Society, London

Fry, R, & Lonsdale, D 1991 *Habitat conservation for insects: a neglected green issue*. The Amateur Entomological Society

Gaffney, V, Thomson, K, & Fitch, S (eds.) 2007 *Mapping Doggerland: the Mesolithic landscapes of the southern North Sea*. Archaeopress, Oxford

Garbutt, A, Burden, A, Maskell, L, Smart, S, Hughes, S, Norris, D & Copper, M 2015 *The status of Habitats Directive Annex I saltmarsh habitats, transition zones and Spartina species in England*. Natural England Commissioned Report, NECR185. Natural England

Gardiner, T, Pilcher, R. & Wade, M 2015 *Sea wall biodiversity handbook*. RPS, Cambridge

Gardiner, S, Hanson, S, Nicholls, R, Zhang, Z, Jude, S, Jones, A, Richards, J, Williams, A, Spencer, T, Cope, S, Gorczynska, M, Bradbury, A, McInnes, R, Ingleby, A & Dalton, H 2007 *The Habitats Directive, coastal habitats and climate change*. Tyndall Centre for Climate Change Research Report Working Paper 108

Gascoyne, A, & Medlycott, M 2014 *Essex historic grazing marsh project*. Essex County Council

George, M 1992 *The landuse, ecology and conservation of Broadland*. Packard, Chichester

Gerard, J 1633 *The herball, or generall historie of plantes* (ed. T Johnson). London

Giles, L 1910 *Sun Tzǔ on the Art of War*. Luzac & Co, London

Gilmour, J S L (ed.) 1972 *Botanical Journeys in Kent and Hampstead*. Hunt Botanical Library, Pittsburgh

Godwin, H 1975 *The history of the British flora* (2nd edition). Cambridge University Press, Cambridge

Gray, S F 1821 *A natural arrangement of British plants*. Baldwin, Cradock & Joy, London

Greenwood, M T, & Wood, P J 2003 Effects of seasonal variation in salinity on a population of *Enochrus bicolor* Fabricius 1792 (Coleoptera: Hydrophilidae) and implications for other beetles of conservation interest. *Aquatic Conservation: Marine and Freshwater Ecosystems* 13(1): 21–34

Grigson, G 1955 *The Englishman's flora*. J M Dent, London

Groves, H, & Groves, J 1882 *On Spartina townsendi. Journal of Botany*, London, 20: 1–20

Groves, J 1913 Henry Groves (1855–1912). *Journal of Botany British and Foreign.* 51: 73–79

Groves, J 1927 The story of our *Spartina. Proceedings of the Isle of Wight Natural History and Archaeological Society* 1(8): 509–513

H.M. Government 2016 *National flood resilience review*. Defra, London

Hall, A R, & Kenward, H K 1990 Environmental evidence from the *Colonia*: General Accident and Rougier Street. *The Archaeology of York* AY 14(6). York Archaeological Trust, York

Hall, A R, Kenward H K, and Williams, D 1980 Environmental evidence from Roman deposits in Skeldergate. *The Archaeology of York* AY 14(3). York Archaeological Trust, York

Hall, A R, Kenward, H K, and Jones, A K G 1986 Environmental evidence from a Roman well and Anglian pits in the legionary fortress. *The Archaeology of York* AY 14 (5). York Archaeological Trust, York

Hammond, C O 1983 *The dragonflies of Great Britain and Ireland* (2nd edition). Harley Books, Colchester

Hannaford, J, Pinn E H, & Diaz, A 2006 The impact of Sika grazing on the vegetation and infauna of Arne saltmarsh. *Marine Pollution Bulletin* 53: 56–63

Hart, C 1999 An estimate of the range and population levels of Fisher's Estuarine Moth (*Gortyna borelii lunata* Freyer), (Lep.: Noctuidae) in Essex, July and October 1996. *British Journal of Entomology and Natural History* 11: 129–138

Hawksford, J E, & Hopkins, I J 2011 *The Flora of Staffordshire*. Staffordshire Wildlife Trust, Stafford

Haynes, T A 2016 *Scottish saltmarsh survey national report*. SNH Commissioned Report No. 786

Hempel-Zawitkowska, J 1968 The influence of desiccation at different air humidities on hatchability of *Triops cancriformis* (Bosc) eggs. *Polskie Archiwum Hydrobiologii* 15(2): 183–189

Hempel-Zawitkowska, J 1969 Hatchability of eggs of *Triops cancriformis* (Bosc) in solutions of chlorides and sulfates. *Polskie Archiwum Hydrobiologii* 16(1): 105–114

Hempel-Zawitkowska, J 1970 The influence of strong ultraviolet radiation on hatchability of *Triops cancriformis* (Bosc) eggs. *Polskie Archiwum Hydrobiologii* 17(4): 483–494

Hendry S J, & Edwards, C E 2012 *Alder woodland. Tree condition assessment survey. Mound Alderwoods* SSSI & SAC. SNH Commissioned Report No.499

Hillman, C G 1981 Macroscopic remains of an estuarine flora. In: McGrail, S (ed) *The Brigg 'raft' and her prehistoric environment*. BAR, Oxford

Holland, R 1850 Rarer plants of Cheshire. *The Phytologist* 3: 863

Horsburg, K, & Horritt, M 2006 The Bristol Channel flood of 1607: reconstruction and analysis. *Weather* 61(10): 272–277

Hough, A, Spencer, C, Lowther, S & Muddiman, S 1999 *Definition of the extent and vertical range of saltmarsh*. EA Technical Report W153. Environment Agency, Bristol

Hounsome, G 2013 *Spartina patens* in West Sussex. *BSBI News* 123: 66–67

House of Lords Journal 4: 20 December 1641

Howson, C M, Chambers, S J, Pye, S E & Ware, F J 2014 *Uist lagoon survey*. SNH Commissioned Report No.787

Hubbard, C E 1954 *Grasses*. Penguin, London

Ingerpuu, N & Sarv, V 2015 Effect of grazing on plant diversity of coastal meadows in Estonia. *Annales Botanica Fennici* 52: 84–92

Ingram, R & Noltie, H J 1981 *The Flora of Angus*. Dundee Museums, Dundee

Izco, J, San Leon, D G & Sanchez, J M 1999 *Spartina patens* as a weed in Galician saltmarshes (NW Iberian Peninsula). *Hydrobiologia* 415: 213–222

Jackson, W 1669 Some inquiries concerning the salt-springs and way of salt-making at Nantwich in Cheshire. *Philosophical Transactions of the Royal Society of London* 4: 1060–1067

Jones, A 1999 *Limosella australis*. In: Wigginton, M J *British Red Data Books 1 Vascular Plants* (3rd edition). JNCC, Peterborough

Kent, D H 1975 *The historical Flora of Middlesex*. Ray Society, London

King, D 2010 *Solent wader and Brent Goose strategy*. HIWWT, Curdridge

Kirby, P 1992 *Habitat management for invertebrates: a practical handbook*. Joint Nature Conservation Committee, Peterborough and RSPB, Sandy, Bedfordshire

Klekowski, R Z & Hempel-Zawitkowska, J 1968 Hatchability of the eggs of *Triops cancriformis* (Bosc) eggs in diluted seawater. *Polskie Archiwum Hydrobiologii* 15(3): 269–277

Lamond, E 1890 *Walter of Henley's Husbandry*. Longmans, Green & Co, New York

Leach, S J 1988 Rediscovery of *Haliminone pedunculata* in Britain. *Watsonia* 17: 170–171

Leary, J 2015 *The remembered land: surviving sea-level rise after the last Ice Age*. Debates in Archaeology. Bloomsbury, London

Lee, J A 1975 The conservation of British inland salt-marshes. *Biological Conservation* 8: 143–5

Lee, J A 1977 The vegetation of British inland salt marshes. *Journal of Ecology* 65: 673–698

Lindley, K 1982 *Fenland riots and the English revolution*. Heinemann, London

Loiseleur-Deslongchamps, J L A 1807 *Flora Gallica*. Lutetiae

Lousely, J E & Kent, D H 1981 *Docks and knotweeds of the British Isles: BSBI Handbook No.3*. BSBI, London

Madgwick, R, Sykes, N, Miller, H, Symmons, R, Morris, J & Lamb, A 2013 Fallow Deer (*Dama dama dama*) management in Roman south-east Britain. *Archaeological and Anthropological Sciences* 5(2): 111–122

Marshall, E S 1914 *A supplement to the Flora of Somerset*. SANHS, Taunton

Merrett, C 1666 *Pinax rerum naturalium britannicarum*. London

Millennium Ecosystem Assessment 2005 *Ecosystem and human well-being: Synthesis*. Island Press, Washington D.C.

Millin, S (ed.) 2010 *UKCIP Briefing Notes. UKCP09 sea level change estimates*. On www.ukcip.org.uk

Ministry of Town and Country Planning 1947 *Conservation of nature in England and Wales*. Cmd. 7122. HMSO, London

Montgomery, W I, Provan, J, McCabe, B M & Yalden, D W 2014 Origin of British and Irish Mammals: disparate post-glacial colonisation and species introductions. *Quaternary Science Reviews* 98: 144–165

Moore, C G, Saunders, G, Mair, J M & Lyndon, A R 2006 *The establishment of site condition monitoring of a saline lagoon. Loch an Duin*. SNH Commissioned Report No.150

Moore, J A 1986 *Charophytes of Great Britain and Ireland: BSBI Handbook No.5*. BSBI, London

Mossman, H L, Davy, A J & Grant, A 2012 Does managed coastal realignment create saltmarshes with 'equivalent biological characteristics' to natural reference sites? *Journal of Applied Ecology* 49(6): 1446–1456

Murray, R P 1896 *The Flora of Somerset*. SANHS, Taunton

Natural Capital Committee 2015 *The state of natural capital – Third report to the Economics Affairs Committee*. On: www.gov.uk/government/uploads/system/uploads/attachment_data/file/516725/ncc-state-natural-capital-third-report.pdf

NCC 1981 *Nature Conservancy Council: 7th Annual Report*. NCC, Peterborough

NE & RSPB 2014 Climate Change Adaptation Manual. On: www.gov.uk/Natural England

Newton, A 1971 *Flora of Cheshire*. Cheshire Community Council, Chester

Nottage, A J & Robertson, P A 2005 *The saltmarsh creation handbook*. RSPB/CIWEM, Sandy

Nowers, J E, & Wells, J G 1892 Notes on a salt-marsh at Branston. *Transactions of the Burton-on-Trent Natural History and Archaeological Society* 2: 50–57

Oliver, F W 1913 Some remarks on Blakeney Point, Norfolk. *Journal of Ecology* 1: 4–15

Paget, C J, & Paget, J 1834 *Sketch of the natural history of Yarmouth etc*. Longman & Co, London

Pallis, M 1911 Salinity in the Norfolk Broads. 1. On the cause of the salinity of the broads of the River Thurne. *Geographical Journal* 37: 284–291

Petch, C P, & Swann, E L 1962 West Norfolk plants today. Supplement to Proc. BSBI 4(4)

Phelan, N, Shaw A, & Baylis, A 2011 *The extent of saltmarsh in England and Wales: 2006–2009*. Environment Agency, Bristol

Pliny the Elder (n.d.) *The Natural History: Book 24*. Translated by Bustock, J & Riley, H T 1855. Taylor and Francis, London

Plot, R 1686 *The natural history of Staffordshire*. Printed at the Theater, Oxford

Postgate, M R 1973 Field systems of East Anglia. In: Baker, A R H, & Butlins, R A (eds.) *Studies of field systems in the British Isles*. Cambridge

Preston, C D 1995 *Pondweeds of Great Britain and Ireland*. BSBI Handbook No.8. BSBI, London

Rand, M, & Mundell, T 2011 *Hampshire rare plant register*. Trollius Publications

Ratcliffe, D A 1977 *A nature conservation review*. Cambridge University Press, Cambridge

Raven, J 1949 Alien plant introductions on the Isle of Rhum. *Nature* 163: 104–105

Ray, J 1660 *Catalogus plantarum circa Cantabrigiam nascentium*. Cambridge

Ray, J 1690 *Synopsis methodica stirpum Britannicarum*. London

Rayner, J F 1907 The conflict of the cord grasses in the Southampton Water: an evolutionary incident. *Proceedings of the Hampshire Field Club and Archaeological Society* 6: 225–229

Rhind, P M 1995 *A review of saltmarsh vegetation surveys in Wales*. Countryside Council for Wales, Bangor

Rich, T C G 1991 *Crucifers of Great Britain and Ireland: BSBI Handbook No.6*. BSBI, London

Riddelsdell, H J, Hedley G, & Price, W R 1948 *Flora of Gloucestershire*. The Cotteswold Naturalists' Field Club, Cheltenham

Ringwood, Z, Hill J, & Gibson, C 2002 Observations on the ovipositing strategy of *Gortyna borelii* Pierret, 1837 (Lepidoptera, Noctuidae) in a British population. *Acta Zoologica Academiae Scientiarum Hungaricae* 48(2): 89–99

Ringwood, Z, Hill J, & Gibson, C 2004 Conservation management of *Gortyna borelii lunata* (Lepidoptera: Noctuidae) in the United Kingdom. *Journal of Insect Conservation* 8: 173–183

Rippon, S 1996 *The Gwent Levels historic landscape study: A report to CADW and CCW*. University of Reading, Reading

Rippon, S 1997 *The Severn estuary: Landscape evolution and wetland reclamation*. LUP, Leicester

Rippon, S 2006 *Landscape, community and colonisation: the north Somerset levels during the 1st and 2nd millennia AD*. Council for British Archaeology, York

Ritchie, J 1949 *Report of the Scottish wild life conservation committee*. Cmnd 7814. HMSO, Stirling

Robinson, D 1987 The Wash: geographical and historical perspectives. In: Doody, P, & Barnett, B (eds.) *The Wash and its environment*. Research and Survey in Nature Conservation no.7. NCC, Peterborough

Rodwell, J S (ed.) 1991 *British Plant Communities. Volume 2: Mires and heaths*. Cambridge University Press, Cambridge

Rodwell, J S (ed.) 1992 *British Plant Communities. Volume 3: Grassland and montane communities*. Cambridge University Press, Cambridge

Rodwell, J S (ed.) 1995 *British Plant Communities. Volume 4: Aquatic Communities, swamps and tall-herb fens*. Cambridge University Press, Cambridge

Rodwell, J S (ed.) 2000 *British Plant Communities. Volume 5: Maritime communities and vegetation of open habitats*. Cambridge University Press, Cambridge

Roe, R G B 1981 *The Flora of Somerset*. SANHS, Taunton

Rothschild, M, & Marren, P 1997 *Rothschild's reserves: Time and fragile nature*. Harley Books, Colchester

Rowe, G, Harris, J D & Beebee, T J C 2006 Lusitania revisited: a phylogeographic analysis of the Natterjack Toad *Bufo calamita* across its entire biogeographical range. *Molecular Phylogenetics and Evolution* 39: 335–346

Rudder, S 1779 *A new history of Gloucestershire*. Rudder, Cirencester

Russell, R C 1968 *The enclosure of Barton-upon-Humber 1793–96*. Workers Educational Association, Barton

Saintilan, N (ed.) 2009 *Australian saltmarsh ecology*. CSIRO Publishing, Victoria

Salisbury, R A 1805 *An account of a storm of salt, which fell in January 1803*. Read 5 February 1805. Manuscript SP984. Library of the Linnean Society of London

Sanderson, N A S 1996, 1997, 2003, 2008 *Lower Test Marshes*. Unpublished reports to Hampshire and Isle of Wight Wildlife Trust

Schlösser, D 1756 Extract d'une lettre de Monsieur le Docteur Schlösser concernant un insect peu connu. *Observations periodiques sur la physique l'histoire naturelle et les beaux arts de Gautier* 58–60

Scott, G A J 1995 *Canada's vegetation: A world perspective*. McGill-Queen Press, MQUP

Shawcross, F W 1960 The excavation of *Bos primigenius* at Lowe's Farm, Littleport. *Proceedings of the Cambridge Antiquarian Society* 54: 3–17

Shoard, M 1980 *The theft of the countryside*. Temple Smith, London

Stace C 2010 *New Flora of the British Isles* (3rd edition). Cambridge University Press, Cambridge

Stamp, D 1942 *The land utilisation survey of Britain: Maps 1 & 2*. Geographical Publications Survey Ltd

Staniforth, R J, Griller, N & Lajzerowicz, C 1998 Soil seed banks from coastal subarctic ecosystems of Bird Cove, Hudson Bay. *Écoscience* 5(2): 241–249

Stanton, D W G, Mulville J A, & Burford, M W 2016 Colonization of the Scottish Islands via long-distance neolithic transport of Red Deer (*Cervus elaphus*). *Proceedings of the Royal Society B* 1283: 1828

Steers, J A 1946 *The coastline of England and Wales*. Cambridge University Press, Cambridge

Sterry, P 1991 British bush-crickets. *British Wildlife* 2(4): 233–237

Stevenson, J 2002 *The benefits to fisheries of UK intertidal saltmarsh areas*. R&D Technical Report E2-061/TR. Environment Agency, Bristol

Stewart, N F 1996 Stoneworts – Connoisseurs of clean water. *British Wildlife* 8(2): 92–99

Stewart, N F 2004 *Important stonewort areas of the United Kingdom*. Plantlife International, Salisbury

Sutherland, W J, & Hill, D A (eds.) 1995 *Managing habitats for conservation*. Cambridge University Press, Cambridge

Tabley, Baron de 1899 *The Flora of Cheshire*. Longmans, London

Tansley, A G (ed.) 1911 *Types of British vegetation*. Cambridge University Press, Cambridge

Tansley, A G 1939 *The British Isles and their vegetation*. Cambridge University Press, Cambridge

Taylor, M 1999 The Wood. In: Pearson, M P & Sharples, N *Between land and sea. Excavations at Dun Vulan. South Uist*. Sheffield Academic, Sheffield

Thornton, D & Kite, D J 1990 *Changes in the extent of Thames estuary grazing marshes*. NCC, Peterborough

Townsend, F 1904 *Flora of Hampshire*. Lovell Reeve & Co, London

Tubbs, C R 1999 *The ecology, conservation and history of the Solent*. Packard, Chichester

UKNEA 2014 *Synthesis of key findings: UK National Ecosystem Assessment*. UNEP-WCMC, LWEC UK

Van de Noort, R 2004 *The Humber wetlands*. Windgather Press, Cheshire

Vera, F W M 2000 *Grazing ecology and forest history*. CABI, Wallingford

WALSC. 1954 *Monmouthshire Moors investigation: draft report*. Welsh Agricultural Land Sub-Commission, Aberystwyth

Wetland Vision Project 2008 *A 50-year vision for wetlands. England's wetland landscape: securing a future for nature, people and the historic environment*. English Heritage/Environment Agency/Natural England/RSPB/The Wildlife Trusts, Sandy

Wilkin, S (ed.) 1835 *Sir Thomas Browne's works*. Pickering, London

Williamson, T 2006 *England's landscape: East Anglia*. English Heritage, London

Wodrow, R 1721 *The history of the sufferings of the Church of Scotland from Restoration to Revolution*. James Watson, Edinburgh

Wyatt, G 2002 Proposed introduction of *Atriplex pedunculata* to a tidally-influenced site on the Essex coast. *BSBI News* 92: 19–22

Yalden, D 1999 *The history of British mammals*. Poyser, London

Yalden, D, & Albarella, U 2009 *The history of British birds*. Oxford University Press, Oxford

Species names

Where standard English names for species exist, these are used throughout the text, with the names of vascular plants following Stace (2010). The list below provides a cross-reference to the current scientific name, with Gaelic or Welsh names in brackets.

Adder's-tongue Spearwort *Ranunculus ophioglossifolius*
Alder *Alnus glutinosa*
Alpine Clubmoss *Diphasiastrum alpinum*
American Pondweed *Potamogeton epihydrus*
American Skunk-cabbage *Lysichiton americanus*
Annual Beard-grass *Polypogon monspeliensis*
Annual Sea-blite *Suaeda maritima*
Arctic Fox *Alopex lagopus*
Arctic Lemming *Dicrostonyx torquatus*
Atlantic Scurvygrass *Cochlearia atlantica*
Aurochs *Bos primigenius*
Avocet *Recurvirostra avosetta*

Babington's Orache *Atriplex glabriuscula*
Baltic Bryum *Bryum marratii*
Baltic Clam *Macoma balthica*
Baltic Rush *Juncus balticus* (Luachair Bhailtigeach)
Baltic Stonewort *Chara baltica*
Barnacle Goose *Branta leucopsis*
Bass *Dicentrarchus labrax*
Bean Goose *Anser fabalis*
Bearded Stonewort *Chara canescens*
Bearded Tit *Panurus biarmicus*
Beaver *Castor fiber*
Bembridge Beetle *Paracymus aeneus*
Bewick's Swan *Cygnus columbianus*
Bird's-foot Clover *Trifolium ornithopodioides*
Bittern *Botaurus stellaris*

Black Bog-rush *Scheonus nigricans* (Sèimhean Dubh)
Black Stork *Ciconia nigra*
Black-headed Gull *Chroicocephalus ridibundus*
Black-winged Stilt *Himantopus himantopus*
Bloody-red Mysid *Hemimysis anomala*
Bog-sedge *Carex limosa*
Borrer's Saltmarsh-grass *Puccinellia fasciculata*
Brackish Water-crowfoot *Ranunculus baudotii*
Brent Goose *Branta bernicula*
Brine Shrimp *Artemia salina*
Brookweed *Samolus valerandi*
Brown Beak-sedge *Rhynchospora fusca*
Brown Bear *Ursus arctos*
Bulbous Foxtail *Alopecurus bulbosus*
Buttonweed *Cotula coronopifolia*

Canada Goose *Branta canadensis*
Celery-leaved Buttercup *Ranunculus sceleratus*
Cetti's Warbler *Cettia cetti*
Chaffweed *Centunculus minima* (Falcair Mìn)
Chilean Flamingo *Phoenicopterus chilensis*
Chinese Mitten Crab *Eriocheir sinensis*
Common Blue Damselfly *Enallagma cyathigerum*
Common Cord-grass *Spartina anglica*
Common Crane *Grus grus*
Common Goldeneye *Bucephala clangula*
Common Knapweed *Centaurea nigra*
Common Nettle *Urtica dioica*
Common Oyster *Ostrea edulis*
Common Reed *Phragmites australis*
Common Saltmarsh-grass *Puccinellia maritima* (Feur Rèisg Ghoirt)
Common Scurvygrass *Cochlearia officinalis* (Am Maraiche)
Common Sea-lavender *Limonium vulgare*

Common Snipe *Gallinago gallinago*
Common Spike-rush *Eleocharis palustris*
Cowbane *Cicuta virosa*
Cowslip *Primula veris*
Creeping Bent *Agrostis stolonifera* (Fioran)
Crested Dog's-tail *Cynosurus cristatus*
Curlew *Numenius arquata*
Curved Sedge *Carex maritima* (Seisg Bheag Dhubh-
cheannach)

Dainty Damselfly *Coenagrion scitulum*
Dalmatian Pelican *Pelecanus crispus*
Dandelions *Taraxacum* spp. (Beàrnan Brìde)
Danish Scurvygrass *Cochlearia danica*
Dark-bellied Brent Goose *Branta bernicla*
ssp *bernicla*
Dark-red Helleborine *Epipactus atrorubens*
De Folin's Lagoon Snail *Caecum armoricum*
Distant Sedge *Carex distans*
Dittander *Lepidium latifolium*
Divided Sedge *Carex divisa*
Domestic Sheep *Ovis aries*
Dotted Sedge *Carex punctata*
Dotterel *Charadrius morinellus*
Dwarf Spike-rush *Eleocharis parvula*
Dwarf Willow *Salix herbacea*

Early Marsh-orchid *Dactylorhiza incarnata*
ssp *incarnata*
Early Orache *Atriplex praecox*
Eelgrasses *Zostera* spp. (Bilearach)
Eight-stamened Waterwort *Elatine hydropiper*
Elk *Alces alces*
English Scurvygrass *Cochlearia anglica*
Essex Emerald *Thetidia smaragdaria*
Estuarine Sedge *Carex recta*
Eurasian Teal *Anas crecca*
Eurasian Wigeon *Anas penelope*
European Bison *Bison bonasus*
European Eel *Anguilla anguilla*
Eyebrights *Euphrasia* spp. (Lus nan Leac)

Fallow Deer *Dama dama*
False Fox-sedge *Carex otrubae*
Fat-hen *Chenopodium album*
Fennel Pondweed *Potamogeton pectinatus*
Fisher's Estuarine Moth *Gortyna borelii*
Fleabane *Pulicaria dysenterica*

Flecked General *Stratiomys singularior*
Floating Pennywort *Hydrocotyle ranunculoides*
Foxtail Stonewort *Lamprothamnium papulosum*
Frosted Orange *Gortyna flavago*

Glassworts *Salicornia* and *Sarcocornia* spp.
Golden Dock *Rumex maritimus*
Golden Plover *Pluvialis apricaria*
Golden-samphire *Inula crithmoides*
Good-King-Henry *Chenopodium bonus-henricus*
Goosander *Mergus merganser*
Goosefoots *Chenopodium* spp.
Grass-leaved Orache *Atriplex littoralis*
Grass-wrack Pondweed *Potamogeton compressus*
Great Silver Water Beetle *Hydrophilus piceus*
Greater Pond-sedge *Carex riparia*
Green-winged Orchid *Anacamptis morio*
Grey Club-rush *Schoenoplectus tabernaemontani*
Grey Heron *Ardea cinerea*
Greylag Goose *Anser anser*

Hair Sedge *Carex capillaris*
Hard-shelled Clam *Mercenaria mercenaria*
Heather *Calluna vulgaris*
Heim's Pottia *Hennediella heimii*
Hemlock Water-dropwort *Oenanthe crocata*
Herring Gull *Larus argentatus*
Hoary Whitlowgrass *Draba incana*
Hog's Fennel *Peucedanum officinale*
Holly-leaved Naiad *Najas marina*
Holy-grass *Hierochloe odorata*
Honewort *Trinia glauca*
Horned Pondweed *Zannichellia palustris*
Horned Wrack *Fucus ceranoides*

Ivell's Sea-anemone *Edwardsia ivelli*
Ivy *Hedera helix*

Kelp *Laminaria* spp.
Killer Shrimp *Dikerogammarus villosus*
Knot *Calidris canutus*
Knotted Clover *Trifolium striatum*

Lagoon Cockle *Cerastoderma glaucum*
Lagoon Sand Shrimp *Gammarus insensibilis*
Lagoon Sandworm *Armandia cirrhosa*
Lagoon Sea Slug *Tenellia adspersa*
Lapwing *Vanellus vanellus*

Lax-flowered Sea-lavender *Limonium humile*
Least Lettuce *Lactuca saligna*
Lesser Black-backed Gull *Larus fuscus*
Lesser Pondweed *Potamogeton pusillus*
Lesser Reed-mace *Typha angustifolia*
Lesser Sea-spurrey *Spergularia marina*
Little Grebe *Tachybaptus ruficollis*
Long-bracted Sedge *Carex extensa*
Long-horned General *Stratiomys longicornis*
Long-stalked Orache *Atriplex longipes*
Lynx *Lynx lynx*

Mallard *Anas platyrhynchos*
Mammoth *Mammuthus primigenius*
Marsh Fleawort *Tephroseris palustris*
    ssp *palustris*
Marsh Foxtail *Alopecurus geniculatus*
Marsh Harrier *Circus aeruginosus*
Marsh Lousewort *Pedicularis palustris*
    (Lus Riabhach)
Marsh Pennywort *Hydrocotyle vulgaris*
    (Lus na Peighinn)
Marsh Sow-thistle *Sonchus palustris*
Marsh-mallow *Althaea officinalis*
Marsh-orchids *Dactylorhiza* spp. (Mogairlean)
Matted Sea-lavender *Limonium bellidifolium*
Meadow Barley *Hordeum secalinum*
Meadowsweet *Filipendula ulmaria*
Mediterranean Gull *Larus melanocephalus*
Meniscus Midge *Dixella attica*
Mistletoe *Viscum album*
Mountain Avens *Dryas octapetala*
Mudwort *Limosella aquatica*
Muscid Fly *Coenosia karli*

Narrow-clawed Crayfish *Astacus leptodactylus*
Natterjack Toad *Epidalea calamita*
New Zealand Pigmyweed *Crassula helmsii*
Northern Pintail *Anas acuta*
Northern River Crangonyctid *Crangonyx pseudogracilis*
Norway Lemming *Lemmus lemmus*

Oblong-leaved Sundew *Drosera intermedia*
Oraches *Atriplex* spp.
Osprey *Pandion haliaetus*
Otter *Lutra lutra*
Oystercatcher *Haematopus ostralegus*

Parsley Water-dropwort *Oenanthe lachenalii*
Pedunculate Sea-purslane *Atriplex pedunculata*
Perennial Beard-grass × *Agropogon littoralis*
Pink-footed Goose *Anser brachyrhynchus*
Pipewort *Eriocaulon aquaticum*
Polar Bear *Ursus maritimus*
Puffin *Fratercula arctica*
Purple Glasswort *Salicornia ramosissima*

Quagga Mussel *Dreissena rosteriformis bugensis*
Quaking-grass *Briza media*

Red Deer *Cervus elaphus*
Red Fescue *Festuca rubra* (Fèisd Ruadh)
Red Goosefoot *Chenopodium rubrum*
Red Swamp Crayfish *Procambarus clarkii*
Redshank *Tringa totanus*
Reed Sweet-grass *Glyceria maxima*
Reed Warbler *Acrocephalus scirpaceus*
Reflexed Saltmarsh-grass *Puccinellia distans*
Reindeer *Rangifer tarandus*
Roe Deer *Capreolus capreolus*
Roesel's Bush-cricket *Metrioptera roeselii*
Rough Stonewort *Chara aspera*
Royal Fern *Osmunda regalis*
Ruff *Philomachus pugnax*

Saiga Antelope *Saiga tatarica*
Saltmarsh Aster *Aster squamatus*
Saltmarsh Flat-sedge *Blysmus rufus* (Seisg Rèisg Ghoirt)
Saltmarsh Goosefoot *Chenopodium chenopodioides*
Saltmarsh Horsefly *Atylotus latistriatus*
Saltmarsh Rush *Juncus gerardii*
Saltmarsh Sedge *Carex salina*
Saltmeadow Cord-grass *Spartina patens*
Scarce Emerald *Lestes dryas*
Scottish Scurvygrass *Cochlearia officinalis* ssp *scotica* (Carran Albannach)
Scottish Primrose *Primula scotica*
Sea Arrowgrass *Triglochin maritimum*
Sea Aster *Aster tripolium*
Sea Aster Bee *Colletes halophilus*
Sea Barley *Hordeum marinum*
Sea Beet *Beta vulgaris* ssp *maritima*
Sea Clover *Trifolium squamosum*
Sea Club-rush *Bolboschoenus maritimus*
Sea Couch *Elytrigia atherica*

Sea Kale *Crambe maritima*
Sea Plantain *Plantago maritima* (Slàn-lus na Mara)
Sea Rush *Juncus maritimus*
Sea Spurge *Euphorbia paralias*
Sea Wormwood *Artemisia maritima*
Sea-heath *Frankenia laevis*
Sea-milkwort *Glaux maritima* (Lus na Saillteachd)
Sea-purslane *Atriplex portulacoides*
Seaside Centaury *Centaurium littorale*
Sharp Rush *Juncus acutus*
Shelduck *Tadorna tadorna*
Shetland Pondweed *Potamogeton rutilus*
Shipworm *Teredo navalis*
Shore Dock *Rumex rupestris*
Short-eared Owl *Asio flammeus*
Shrill Carder Bee *Bombus sylvarum*
Shrubby Orache *Atriplex halimus*
Shrubby Sea-blite *Suaeda vera*
Siberian Brown Lemming *Lemmus sibiricus*
Sika Deer *Cervus nippon*
Silverweed *Potentilla anserina* (Brisgean)
Slender Centaury *Centaurium tenuiflorum*
Slender Hare's-ear *Bupleurum tenuissimum*
Slender Spike-rush *Eleocharis uniglumis*
Slender-leaved Pondweed *Potamogeton filiformis*
Small Cord-grass *Spartina maritima*
Small Water-pepper *Persicaria minor*
Small-fruited Yellow-sedge *Carex oederi* (Seisg nam
    Measan Beaga)
Smew *Mergus albellus*
Smooth Cord-grass *Spartina alterniflora*
Somerset Hair-grass *Koeleria vallesiana*
Southern Migrant Hawker *Aeshna affinis*
Spear-leaved Orache *Atriplex prostrata*
Spiked Water-milfoil *Myriophyllum spicatum*
Spiny-cheek Crayfish *Orconectes limosus*
Spoonbill *Platalea leucorodia*
Stag Beetle *Lucanus cervus*
Starlet Sea-anemone *Nematostella vectensis*
Starling *Sturnus vulgaris*
Steppe Bison *Bison pricus*
Sticky Catchfly *Silene viscaria*
Stiff Saltmarsh-grass *Puccinellia rupestris*
Strawberry Clover *Trifolium fragiferum*

Subterranean Clover *Trifolium subterraneum*
Summer-cypress *Bassia scoparia*
Sunflower *Helianthus annuus*
Sweet Vernal-grass *Anthoxanthum odoratum*

Tadpole Shrimp *Triops cancriformis*
Tamarisk *Tamarix gallica*
Tasselweeds *Ruppia* spp.
Tasteless Water-pepper *Persicaria mitis*
Tentacled Lagoon-worm *Alkmaria romijni*
Thrift *Armeria maritima* (Neòinean Cladaich)
Toad Rush *Juncus bufonius*
Townsend's Cord-grass *Spartina* × *townsendii*
Trailing Azalea *Kalmia procumbens*
Tree Groundsel *Baccharis halimifolia*
Trembling Sea-mat *Victorella pavida*
Triangular Club-rush *Schoenoplectus triqueter*

Upright Chickweed *Moenchia erecta*
Urial *Ovis orientalis*

Wall *Lasiommata megera*
Water Fern *Azolla filiculoides*
Water-milfoils *Myriophyllum* spp. (Snàthainn
    Bhàthaidh)
Welsh Mudwort *Limosella australis* (Lleidlys
    Cymreig)
White Stork *Ciconia ciconia*
White-fronted Goose *Anser albifrons*
White-tailed Eagle *Haliaeetus albicilla*
Whooper Swan *Cygnus cygnus*
Wild Boar *Sus scro*
Wild Celery *Apium graveolens*
Wild Horse *Equus ferus*
Wild Teasel *Dipsacus fullonum*
Wolf *Canis lupus*
Woolly Rhinoceros *Coelodonta antiquitatis*

Yellow Horned-poppy *Glaucium flavum*
Yellow Iris *Iris pseudacorus*
Yellow Wagtail *Motacilla flava*
Yew *Taxus baccata*

Zebra Mussel *Dreissena polymorpha*

Illustration credits

All photographs are © the author, except for those listed below. Bloomsbury Publishing would like to thank those listed below for providing photographs and for permission to reproduce copyright material within this book. Whilst every effort has been made to trace and acknowledge all copyright holders, we would like to apologise for any errors or omissions, and invite readers to inform us so that corrections can be made in any future editions.

1 © Paul Hobson/Nature Picture Library; 2-3 © Ernie Janes/Nature Picture Library; 6 © Sixpixx/Shutterstock.com; 12 © David Clapp/Getty Images; 17 © Lorne Gill/Scottish Natural Heritage; 22 © Patrick Aventurier/Getty Images; 23 top © Ernie James/Nature Picture Library; 24 © Steven Falk; 27 top © Ernie James/Nature Picture Library, bottom © Special Collections Research Center University of Chicago; 34 © viledevil/iStock; 36 © David Wain/Staffordshire Wildlife Trust; 37 © Bob Gibbons; 39 © Steven Falk; 42 inset © Lliam Rooney; 45 bottom © D.Haaksma/Bryopix; 47 left © Lliam Rooney, right © Henri Koskinen/Shutterstock.com; 48 © Universal History Library/Getty Images; 49 © duncan1890/iStock; 54 top © StockSolutions/iStock; 57 top © Navapon Plodprong/iStock, bottom © Alex Hyde/Nature Picture Library; 60 © Lisland/iStock; 61 top © Lorne Gill/Scottish Natural Heritage, bottom © Andrew Diack; 62 inset © Dave Dunford, main picture © arrowsg/iStock; 63 inset © Lliam Rooney; 64 bottom © Steven Falk; 65 main picture © Philip Halling; 73 bottom © George Logan/Scottish Natural Heritage; 74 top © Jeremy Roberts; 78 © Mike Baldwin;

79 © Stephen Dorey/Getty Images; 80 © Mark Heighes/HIWWT; 81 bottom © Joe Dunckley/iStock; 84 © Lorne Gill/Scottish Natural Heritage; 88,90 & 91 top © Bob Gibbons; 91 bottom © S J Allen/Shutterstock.com; 95 © scubaluna/Shutterstock.com; 99 © Bob Gibbons; 101 © Tom Langton/HCI; 106 © StockSolutions/Alamy; 110 © David Tipling/birdphoto.co.uk; 111 © Ben Cranke/Getty Images; 113 © blickwinkel/Blumenstein/Alamy; 117 © Jan Baks/NiS/Minden Pictures/Getty Images; 119 © Yva Momatiuk & John Eastcott/Minden Pictures/Getty Images; 121 © Bob Gibbons; 123 bottom © Heartland Arts/Shutterstock.com; 126 © ultraforma/iStock; 128 top © Bob Gibbons; 129 © SSPL/Getty Images; 131 © hmproudlove/iStock; 134 © Steven Falk; 135 © Graham Eaton/Nature Picture Library; 136 © Global Warming Images/Alamy; 140 © Ethan Welty/Getty Images; 146 top © Alex Hyde/Nature Picture Library; 149 top © PhilMacD Photography/Shutterstock.com, bottom © Bob Gibbons; 150 © Hulton Archive/Getty Images; 155 © Lliam Rooney; 157 © Stephen Dorey/Getty Images; 158 © Nick Upton/Nature Picture Library; 159 © Lliam Rooney; 161 bottom © Mike Boyland/iStock; 162 top © Roger Key, bottom © Steven Falk; 163 © Steven Falk; 165 © Gideon Chilton/Getty Images; 170 © David Woodfall/Nature Picture Library; 172 © Robert Canis/Getty Images; 175 © Terry Whittaker/2020 Vision/Nature Picture Library; 176 © Ian Rose/FLPA; 178 © Buiten-Beeld/Alamy; 183 bottom © Lliam Rooney; 185 © Steven Falk; 187 © Saxifraga/Peter Meininger; 188 © Ferdinando Valverde/Getty Images; 189 © Popperfoto/Getty Images; 192 © Kim Taylor/Nature Picture Library;

193 © BasPhoto/Shutterstock.com; 194 top © Steven Falk, bottom © Rob Aguilar/Smithsonian Environmental Research Center; 195 top © Jim Champion, bottom kodachrome25/iStock; 197 left © Steven Falk, right © Louis Falk; 198 © Paul Sterry/Nature Photographers Ltd; 209 © Journal of Botany; 218 © Geoff Hounsome; 220 © David Tipling/birdphoto.co.uk; 223 © SSPL/Getty Images; 226 top © Veselin Gramatikov/Shutterstock.com, bottom © Florilegius/SSPL/Getty Images; 227 top © Daniel Petrescu/Shutterstock.com, bottom © Montipaiton/Shutterstock.com; 228 top © David Tipling/birdphoto.co.uk, bottom © Heritage Images/Getty Images; 232 © Last Refuge/Robert Harding/Getty Images; 234 © Hulton Archive/Getty Images; 241 © Roger Key; 242 © Andy Hay/RSPB Images; 251 © Gordon Langsbury/RSPB Images; 254 top © Stafford Borough Council; 269 © Chris Gomersall/RSPB Images; 270 © Universal Images Group/Getty Images; 271 © Universal History Archive/Getty Images; 273 inset © Lliam Rooney;

275 © Mantonature/iStock; 277 © Helen Hotson/Shutterstock.com; 278 © blickwinkel/Alamy; 279 © Pallis/Vlasto Archive; 285 top © Jenny Hibbert/RSPB Images; 286 © James Osmond/Getty Images; 288 left © Getty Images, right PhotoQuest/Getty Images; 296 © Catherine Chatters; 300 © pengpeng/iStock; 301 © Matteo photos/Shutterstock.com; 303 © Jim Richardson/Getty Images; 304 © Nature Picture Library/Getty Images; 307 © Steven Falk; 308 © David Tipling/birdphoto.co.uk; 310 © Ian Rose/FLPA; 314 © Andrew Britland/HIWWT; 316 © Andy Hay/RSPB Images; 318 © Andre Anita/ iStock; 320 © Ian Cameron-Reid/HIWWT; 321 © Steve Page/HIWWT; 323 © Mark Heighes/HIWWT; 326 © Lliam Rooney; 328 © David Tipling/birdphoto.co.uk; 335 © David Phillips/HIWWT; 336 © Ashley Cooper pics/Alamy Stock Photo ; 337 © Terry Whittaker/2020 Vision/Nature Picture Library; 339 main picture © Peter Macdiamid/Getty Images, inset Dan Kitwood/Getty Images; 340 © I Wei Huang/Shutterstock.com.

Index

Page numbers in **bold** refer to illustrations.

Aberffraw, Anglesey 235
acidity 95, 100
Acrocephalus scirpaceus 335–336,
   370
Add River, Argyll and Bute 80, 82
Aeshna affinis 189, 371
agricultural use 15; *see also* grazing,
   saltmeadow
   common rights 221–222, 228–
   231, 238
   Halvergate debate 280–285
   hay meadows 154, **154**
   leading to loss of marsh 39–40,
   158, **225**, 241, 262–263; *see
   also* drainage of marshes
   machair **94**, 103–105, **104**, **105**
*Agriculture, the Triumph and the
   Shame* (Body) 267
Agropogon littoralis 211, 370
Agrostis stolonifera 90, 153, 156,
   305, 369
Alces alces 22, 118–120, **119**, 142,
   369
Alde estuary, Suffolk 316
Alder 76–78, **78**, 80, 135, 305, 368
Aldersey, Cheshire 352
Alkborough, Lincolnshire **136**,
   137–138, 335–336, **336**
Alkmaria romijni 291, 371
Alnus glutinosa 76–78, **78**, 80, 135,
   305, 368
Alopecurus bulbosus 149, 154–155,
   155, 171, 196, 325, 368
   geniculatus 155, 370
Alopex lagopus 318, 368
Althaea officinalis 174, 223–224,
   224, 272, 285, 370
Am Maraiche 54–55, **54**, 61, 69–70,
   368
Anacamptis morio 326, 369
Anas acuta 269, 370
   crecca 226, **226**, 269, 369
   penelope 143, 169, 226, 281, 284,
   369
   platyrhynchos 143, 370
Ancholme River 146, 147, **147**
Anderton Nature Park, Cheshire

52, 353
Andrews, Joseph 178
Anguilla anguilla 279, 369
Anser albifrons 281, 371
   anser 90–91, **91**, 143, 369
   brachyrhynchus 281, 284, 370
   fabalis 281, 368
Antelope, Saiga 141, 370
Anthoxanthum odoratum 122, 156,
   371
Apium graveolens 38, 39, **39**, 40, 41,
   47, 175, 178, 326, 371
apothecaries 173–179
archaeophytes 203
Ardea cinerea 268, 369
Armandia cirrhosa 291, 369
Armeria maritima see Thrift
Arne RSPB Nature Reserve, Dorset
   256
Arnold's Marsh, Norfolk **342**
Arrowgrass, Sea 370
   Brigg boat find, Humber 147
   cliffs 60
   inland saltmarshes 35, 39, 41, 46
   Pennine moors 56
   Roman York 148
   Solent 203
   White Sea, Russia 140
arrowgrasses 320
Artemia salina 191–193, **192**, 199,
   368
Artemisia maritima 127, **128**, 171,
   174, 240, 271, 306, 371
Asio flammeus 97, 334, **335**, 371
Astacus leptodactylus 300, 370
Aster 299
   squamatus 299, 370
   tripolium see Aster, Sea
Aster, Saltmarsh 299, 370
   Sea 370
   Blakeney Point 27
   the Broads 272, 281
   inland saltmarshes 36–37, **37**,
   38, 47, 52, 53
   invertebrate foodplant 306, **307**
   Keyhaven marshes 196
   Medmerry 334

Roman York 148
Severn Levels 153, 157
the Solent 207
Titchwell RSPB Nature Reserve
   315
the Wash 223–224
White Sea, Russia 140
Widewater Lagoon **290**
Astonfields LNR, Staffordshire **254**,
   350
Atriplex 56, 186, 298, 370
   glabriuscula 368
   halimus 299, 371
   littoralis 57, 369
   longipes 86, 370
   pedunculata 186–188, **187**, 271,
   291, 305, 370
   portulacoides see Sea-purslane
   praecox 86, 369
   prostrata 45, 86, 371
Atylotus latistriatus **194**, 370
Audlem, Cheshire 352
Aurochs 22, **22**, 118–120, 142,
   223, 368
Avens, Mountain 71, 86, **86**, 370
Avocet **34**, 43–44, 269, 313, 315,
   368
Avon Wildlife Trust 159, 170
Axe River, Somerset **168**, 170–171,
   317, **317**
Aylburton Warth, Gloucestershire
   153, 168–169
Azalea, Trailing 71, 371
Azolla filiculoides 298, 371

Babington, Charles 224
Baccharis halimifolia 299, 371
Bae Ceredigion
   Cantref-y-Gwaelod 134–135
   climatic meeting zone 127–134
   Mawddach estuary **124**
   Morfa Harlech 125–127, **126**
   Nature Conservation Review
   sites 248
   salinity 129
   submerged woodlands 134–135,
   135

Traeth Mawr 130, **131**
Baile Sear, Outer Hebrides 101–102
Balfour-Browne, Frank 114–115
Baring-Gould, Reverend Sabine 48, **48**, 50, 181
Barley, Meadow 281, 370
Sea 224, **225**, 272, 285, **285**, 334, 370
barrier-connected marshes 29
Barton-Upon-Humber, Lincolnshire 223
baselines, shifting 341
Bass 214, 299, 368
Bassia scoparia 56–57, **57**, 299, 371
Beak-sedge, Brown 84, 368
Bear, Brown 142, 368
Polar 112, 370
Beard-grass, Annual 319, **319**, 368
Perennial 211, 370
Beàrnan Brìde 71, 369
Beaulieu River, Hampshire **30**, 214, 304
Beaver 142, 368
Bee, Sea Aster **307**, 370
Shrill Carder 163–164, **163**, 371
Beet, Sea 41, 65, 370
Beetle, Bembridge 291, 368
Great Silver Water 160, 162, **162**, 369
Stag 162, 371
beetles, aquatic 160
Bennett, Arthur 276
Bent, Creeping 90, 153, 156, 305, 369
Beta vulgaris ssp *maritima* 41, 65, 370
Bettyhill, Highlands 72–73, **73**
Bilearach 27, 68, 92, 102, 214, 217, 319, 369
biodiversity 17, 24, 95, 258, 307, 313
Bison bonasus 123, 369
pricus 46, 371
Bison, European 123, 369
Steppe 46, 371
Bittern 226, **227**, 268, 313, 368
Black Book of Carmarthen 134
Blakeney Point, Norfolk 26–27, **27**, **28**, 29, **232**, 233–234, 236, 238–239, 257
Bleadon Levels, Somerset 171
Blysmus rufus 46, 61, 69, **69**, 82, 370
Boar, Wild 142, 371
Bog-rush, Black 70, **70**, 368
Bog-sedge 84, 368
Bolboschoenus maritimus see Club-rush, Sea
Bombus sylvarum 163–164, **163**, 371
Bonn Convention 288

Borth, Ceredigion 135, **135**
Bos primigenius 22, **22**, 118–120, 142, 223, 368
Botaurus stellaris 226, **227**, 268, 313, 368
Brancaster, Norfolk **18**, 238, 239–240
Branston, Staffordshire 46–48, **47**, 350–351
Branta bernicla 269, 368
canadensis 44, 297–298, 368
leucopsis 109–112, **110**, **111**, 368
ssp bernicla 317–322, **318**, **319**, **320**, **321**, 369
Braunton Burrows, Devon 235
Brisgean 62, **88**, 90–92, **91**, 371
British Isles and their Vegetation, The (Tansley) 28
British Plant Communities (Rodwell, ed) 31–32
Briza media 326–327, 370
Broads Authority 282–283, 285
Broads, the
Bennett's accounts 276–278
Browne's accounts 268–270
drainage 280–281
Halvergate debate 280–285
invasive species 301
Paget brothers' accounts 270–272, 273–276
salinity 278–280, 281
Bromfield, Reverend William 204, 205–206, **205**
Bronze Age 23, 118, 146–148, 223, 278
Brookweed **134**, 368
Bae Ceredigion 134
inland saltmarshes 40, 41, 43, 47, 48, 53
perched saltmarshes 62
in Pliny 133
the Solent 298
storm beach saltmarshes 65
Brown, Robert 129–130, **129**
Browne, Sir Thomas 268–270, **270**
brownfield sites 164, 199; see also Keyhaven marshes, Hampshire
Bryum, Baltic 128–129, 368
Bryum marrattii 128–129, 368
BSBI News 55, 56, 218
Bucephala clangula 226, 368
Bupleurum tenuissimum 42–43, **42**, 153, 319, 371
Burgh Marsh, Cumbria 107, **107**, **108**
Burgh Marshes, Norfolk 271
Bury Marsh, Hampshire 215, **216**
Bush-cricket 196–197, **197**
Roesel's 196, 197, 370

Buttercup, Celery-leaved 45, **45**, 298, 368
Buttonweed 297–298, 368
Buxton, Anthony 279

Caecum armoricum 291, 369
Caerlaverock Castle, Dumfries and Galloway **106**
Calidris canutus 226, 337, 369
Calluna vulgaris 82, 369
Calshot, Hampshire **235**
Canis lupus 142, 371
Capreolus capreolus 22, 142, 370
carbon sequestration 294
Cardiff 20, **21**, 167
Cardigan Bay see Bae Ceredigion
Carex capillaris 86, 369
distans 38, 39, 40, 41, 56, 157, 369
divisa 157, 271, 326, **326**, 369
extensa 65, 370
limosa 84, 368
maritima 55, 61, 72–73, **72**, **73**, 305, 369
oederi 61, 69, **69**, 73, 371
otrubae 65, 281, 369
punctata 196, 369
recta 76, 369
riparia 326, 369
salina 73–76, **74**, 257, 370
Carran Albannach 69–70, 370
Castor fiber 142, 368
Catchfly, Sticky 121, **121**, 371
Celery, Wild 38, 39, **39**, 40, 41, 47, 175, 178, 326, 371
Centaurea nigra 156, 163, 368
Centaurium littorale 83, **84**, 371
tenuiflorum 64, **64**, 291, 371
Centaury, Seaside 83, **84**, 371
Slender 64, **64**, 291, 371
Centunculus minima 70, 368
Cerastoderma glaucum 102, 369
Cervus elaphus 22, 142, 304, 370
nippon 303–304, **304**, 371
Cettia cetti 325, 368
Chaffweed 70, 368
Chapman, Val 26–27
Chara 277, 278
aspera 98, 100, 101, 102, 277, 370
baltica 101–102, 277, 368
canescens 101, **101**, 277, 291, 368
Charadrius morinellus 226, **227**, 369
Charles I 222, 228
Charophyta spp. 53, 99–102
Cheltenham, Gloucestershire 40, 41
Chenopodium 56, 298, 369
album 184–185, 186, 207, 369
bonus-henricus 186, 369

chenopodioides 184–185, **185**, 370
  rubrum 184–185, 315, 370
chenopods 184–187, **185**, **187**
Chetwynd, Walter 36
Chichester Harbour, West Sussex 218–219, **218**, 299
Chickweed, Upright 196, 371
Chroicocephalus ridibundus 44, 198, **198**, 368
Ciconia ciconia 143, 269, 271, 371
  nigra 143, 368
Cicuta virosa 276, 369
Circus aeruginosus 313, 335, 370
Clam, Baltic 337, 368
  Hard-shelled 299–300, 369
clay pits 101
Cledford, Cheshire 354
Cley, Norfolk 236, **239**
cliff pools 63–64, **64**
cliff-top vegetation 60–61, **60**
Clifford, Arthur 37–38
Clifford, Thomas 37–38
climate change 20, **21**, 167, 266, 294
climate, oceanic influences on 21–22
climatic zones 21
Clover 110, 157, 163, 196
  Bird's-foot 196, 368
  Knotted 196, 369
  Sea 64, 157, **159**, 171, 178, 272, 319, 370
  Strawberry 40, 148, 153, 171, 174, 371
  Subterranean 196, 371
Club-rush, Grey 38, 41, 45, 48, 53, 369
  Sea 370
    Brigg boat find, Humber 147
    the Broads 281
    inland saltmarshes 38, 41, **41**, 43, 47
    perched saltmarshes 62
    the Solent 207, 297
    Thames estuary 174, 178
    the Wash 223–224
    White Sea, Russia 140
    Triangular 178–179, **178**, 291, 305, 371
Clubmoss, Alpine 71, 368
coastal engineering 16; *see also* coastal realignment projects; sea defences
  biodiversity, impact on 332–333
  drainage of marshes *see* drainage of marshes
  embankments *see* embankments
  flexibility 338
  modifying 167, 313–315, **316**

tidal sluices 76–77, **77**
coastal realignment projects
  Alkborough Flats 335–336, **336**
  Colne **332**
  Cromarty Firth 336–337, **337**
  ecological effects 332–333
  flexibility 338
  legal obligations 332
  livestock grazing 333, 338
  Titchwell **316**
  Wallasea 338–341, **339**, **340**
Cochlearia 54, 69–70
  anglica 153, 169, 177–178, 271, 369
  atlantica 69–70, 368
  danica 54, 57, **57**, 369
  officinalis 54–55, **54**, 61, 69–70, 368
  officinalis ssp *scotica* 69–70, 370
Cockle, Lagoon 102, 369
Coelodonta antiquitatis 46, 276, 371
Coenagrion scitulum 188–189, **188**, 369
Coenosia karli **24**, 370
Colletes halophilus **307**, 370
common rights 129, 221–222, 228–231, 238
Compton Bay, Isle of Wight **58**
Conference on Nature Preservation in Post-war Reconstruction 243–244
Conocephalus spp. 196–197, **197**
Conon estuary, Highlands 77
conservation
  Blakeney Point **232**, 233–234
  Brent Geese 317–322, **318**, **319**, **320**, **321**
  Brexit questions 292–293
  the Broads 280–285
  ecosystem services 293–295
  Fisher's Estuarine Moth 309–313, **310**, **311**
  Gedney Drove End **260**, 263–268, **264**
  Hazelwood Marshes 316
  international treaties and conventions 287–293
  Lower Test Nature Reserve 323–327, **323**, **324**, **325**, **326**
  military sites 259, **259**
  North Norfolk Coast 236–237, 238
  Ray Island 240
  Romney Marshes 240–241
  site and species protection 288–293
  Society for the Promotion of Nature Reserves list 234–235, 237

statutory sector 244, 247, 251–253
  Titchwell **308**, 313–315, **314**, **316**
  voluntary sector 255–259
  Wildlife and Countryside Act (1981) 267–268
Conservation Areas 244, 245–246, 247
Convention for Protection of the Marine Environment of the North East Atlantic 288
Cord-grass, Common **200**, **202**, **208**, 368
  coastal realignment projects 334
  land conversion use 201–202
  origins 201, 207–208
  Severn 165
  the Solent 217–218, **217**
  spread 201–203, 212, 305
  Wallasea 339
  Saltmeadow 218, **218**, 299, 370
  Small **204**, **205**, 371
    Brigg boat find, Humber 147, 203
    decline 214–215, 305
    description 203–204
    distribution 204
    hybridisation 201
    Isle of Wight 215
    the Solent 207, 211–212, 215
    Thames estuary 174
  Smooth 201, 204–207, **205**, **206**, 211–212, 215–216, **216**, 371
  Townsend's 28, 201, 207–208, 210–211, 212, **216**, 217, 371
Cotula coronopifolia 297–298, 368
Couch, Sea 176, 305, 334, **335**, 370
Cowbane 276, 369
Cowslip 327, 369
Crab, Chinese Mitten 300, **300**, 368
Crambe maritima 174, 371
Crane, Common 143, 226, 268, **269**, 368
Crangonyctid, Northern River 301, 370
Crangonyx pseudogracilis 301, 370
Crassula helmsii 298, 370
Crayfish, Narrow-clawed 300, 370
  Red Swamp 300, **301**, 370
  Spiny-cheek 300, 371
Cree River, Dumfries and Galloway 118, 123, 253
creeks 31
crickets, conehead 196–197, **197**
Crinan Ferry, Argyll and Bute 80, **81**, **82**
Cromarty Firth, Highlands 77, 132, 336–337, **337**

Cromwell, Oliver 231
Crown estates 222–223, 230
Curlew 169, 337, **337**, 369
Cwm Ivy, Gower, Glamorgan 330,
    331, 332
Cygnus columbianus 269, 281, 368
    cygnus 143, 371
Cynosurus cristatus 156, 369

Dactylorhiza 52, 90, 332, 370
    incarnate ssp *incarnate* 327, 369
Dama dama 303, 369
Damselfly, Common Blue **285**, 368
    Dainty 188–189, **188**, 369
dandelions 71, 369
Dandy, J E 96
Darby, Henry Clifford 263
Darwin, Charles 191–192
decline of saltmarshes 24, 341
Dee estuary 62, **62**, 255
Deer, Fallow 303, 369
    Red 22, 142, 304, 370
    Roe 22, 142, 370
    Sika 303–304, **304**, 371
Defford Common, Worcestershire
    42–43, 348
definitions of "saltmarsh" 15
Denge Peninsula, Essex 29
Dennes, George Edgar 211–212
Descripto Itineris (Johnson) 173, 177
Dicentrarchus labrax 214, 299, 368
Dickens, Charles 179–180
Dicrostonyx torquatus 318, 368
Dictionnaire oeconomique (Chomel)
    15
Dikerogammarus villosus 301, 369
dilution 16, 21
Diphasiastrum alpinum 71, 368
Dipsacus fullonum 164, 371
distribution maps **14**, **346**
Ditiscus circumflexus 160
Dittander 38, 272, **273**, 369
Dixella attica 162, 370
Dock, Golden 47, 48, 369
    Shore 64–65, **65**, 290, 291, 305,
        371
Doggerland 138–139
    flora and fauna 139–141, **140**,
        142–143
    herbivory impacts 142–143
    human occupation 143–144
    submersion 144–145
    warming period 141–144
Dog's-tail, Crested 156, 369
Dolemoors, Somerset 159–160
Donna Nook, Lincolnshire **259**
Dornoch Firth, Highlands 235, 259
Dotterel 226, **227**, 369
Draba incana 46, 369

drainage of marshes
    Ancholme **147**
    the Broads 270, 280–285
    for crops 158, **158**, 228, **229**,
        230–231, 262–263, 280–285
    Halvergate debate 280–285
    Humber 230-231
    Ingestre Park 37–38
    Leadon and Ell valleys 41
    Romney Marshes 241
    for salt industry 37–38
    the Wash **225**, 228, **229**, 262–
        263
Drayton, Michael 225–226, **226**
Dreissena polymorpha 300, 371
    rosteriformis bugensis 300, 370
driftwood 98–99, **99**
Droitwich, Worcestershire 43,
    348–349
Drosera intermedia 84, 370
Druce, George Claridge 39
Dryas octapetala 71, 86, **86**, 370
Dùn Vùlan broch, South Uist 98
Dyfi estuary, Ceredigion 28, 135,
    252
dynamism 17, 313, 343

Eagle, White-tailed 269, 271, 371
ecology, 20th-century 26–29
ecology, 21st-century 29–33
ecosystem services 293–295
Edinburgh 20, **21**
Edward I 107, **107**, 125
Edwardsia ivelli 290, 291, 369
Eel, European 279, 369
eelgrasses 27, 68, 92, 102, 214,
    217, 319, 369
Elatine hydropiper 147–148, 369
Eleocharis palustris 148, 369
    parvula 132, **133**, 291, 369
    uniglumis 140, 371
Elk 22, 118–120, **119**, 142, 369
Ell River 41
Elmley Marshes, Kent **175**
Elmore, Gloucestershire 156, **157**
Elytrigia atherica 176, 305, 334, **335**,
    370
embankments
    the Broads 270, 280, 281
    Cromarty Firth 336
    the Gower 330, **331**
    Severn 156, 165, 166, 167, 170
    the Solent 193, 197–198, 319,
        321, 322
    Thames estuary 182, 184
    Wallasea 338
    the Wash 186, 222–223, 228–
        229, 262–263, 266, 267
Emerald moth, Essex 240, **241**, 369

Scarce 189, 370
Enallagma cyathigerum **285**, 368
enclosures
    the Broads 270, 280
    Defford Common 43
    loss of common rights 221–222,
        222–223, 228–231, 238
    Severn 154, 160, 221–222
    Thames estuary 182
    the Wash 228–231, **229**, 263
endangered species 305
Enochrus bicolor 44
Environment Agency 15, 167, 170,
    259, 294–295, 327, 333, 335,
    338
Epidalea calamita 115, 116–117,
    117, 118, 120, 291, 370
Epipactus atrorubens 71, 369
Equus ferus 23, 141, 142–143, 371
Eriocaulon aquaticum 84, 370
Eriocheir sinensis 300, **300**, 368
Essex Wildlife Trust 240, 258, **332**
estuarine marshes 29, **30**
etymology of "saltmarsh" 13, 15
Euphorbia paralias 174, 371
Euphrasia 71, 84, 210, 369
    foulaensis 210
    heslop-harrisonii 85
European Union directives 289–
    290, 292
evaporation 16
eyebrights 71, 84, 210, 369

Farlington Marshes, Hampshire **318**,
    319–320, **319**, **320**, 322
Fat-hen 184–185, 186, 207, 369
Fèisd Ruadh 28, 60, 68, 153, 156,
    305, 337, 370
Fennel, Hog's 175–176, **176**, 311–
    312, **311**, 369
Fern, Royal 196, 370
    Water 298, 371
Fescue, Red 28, 60, 68, 153, 156,
    305, 337, 370
Festuca rubra 28, 60, 68, 153, 156,
    305, 337, 370
Feur Rèisg Ghoirt 28, 49, 68, 126,
    153, 239, 305, 315, 337, 368
Field, John 178
Filipendula ulmaria 80, 370
Fingringhoe Wick, Essex **182**, **293**
Fisher, Ben 309
Flamingo, Chilean 191, 368
Flat-sedge, Saltmarsh 46, 61, 69, **69**,
    82, 370
Fleabane 163, 369
Fleawort, Marsh 272, 274–276,
    275, 370
Fleet River, Highlands 77

floods, disastrous **149**, **150**, 151–152, 167
Flora of Berkshire (Druce) 39
Flora of Cambridgeshire (Babington) 224
Flora of Hampshire (Townsend) 205–206
Flora of Norfolk (Beckett and Bull) 276
Flora of Somerset, The (Roe) 158
Flora Tixalliana (Clifford and Clifford) 37–38
Fly, Muscid **24**, 370
Fobbing Marsh, Essex 183, **183**, 184
foreland marshes 29
Fox, Arctic 318, 368
Fox-sedge, False 65, 281, 369
Foxtail, Bulbous 149, 154–155, **155**, 171, 196, 325, 368
  Marsh 155, 370
fragmentation of habitat 341
Frankenia laevis 27, 63, **63**, 224, **290**, 334, 371
Fratercula arctica **60**, 226, 370
Fucus ceranoides 102, 369
  muscoides 67
future 341, 343–344

Gallinago gallinago 271, 281, 369
Gammarus insensibilis 194, 291, 369
Gedney Drove End, Lincolnshire **260**, 263–268, **264**
General soldierfly, Flecked 162, **162**, 369
  Long-horned 64, **64**, 370
geology, solid 18–21
geomorphology 29, 31
Gerald of Wales 134
glacial periods 18–19, 45, 85, 116; see also Doggerland
Glasswort, Purple 52, 370
glasassworts 25–26, **25**, 369
  absence from Highlands 68
  Cambridgeshire fens 278
  deer herbivory 303–304
  description 25–26
  Gower peninsula, Glamorgan 332
  inland saltmarshes 52
  Keyhaven marshes 196
  Lower Test 324
  North Norfolk Coast 27
  Severn 153
  Thames estuary 174
  Titchwell 315
Glastonbury Abbey, Somerset 221–222
Glaucium flavum 174, **174**, 371
Glaux maritima see Sea-milkwort
Glyceria maxima 326, 370

godwits 226
Goethe, Johann Wolfgang von 24
Golden-samphire 127, 174, 369
Goldeneye, Common 226, 368
Good-King-Henry 186, 369
Goosander 226, 369
Goose, Barnacle 109–112, **110**, **111**, 368
  Bean 281, 368
  Brent 269, 368
  Canada 44, 297–298, 368
  Dark-bellied Brent 317–322, **318**, **319**, **320**, **321**, 369
  Greylag 90–91, **91**, 143, 369
  Pink-footed 281, 284, 370
  White-fronted 281, 371
Goosefoot 56, 298, 369
  Red 184–185, 315, 370
  Saltmarsh 184–185, **185**, 370
Gortyna borelii 291, 309–313, **310**, **311**, 369
  flavago 369
Gower peninsula, Glamorgan **32**, 129, 258, **258**, 330, **331**, 332
Grasses (Hubbard) 207–208
grazing, saltmeadow; see also herbivory
  Bae Ceredigion 129, 132
  benefits 123
  the Broads 280–285
  consequences of no grazing 109, 209, 305–306, 307
  early writings on 15
  ecological benefits 23–24, 42, 45, 83, 84–85, 104–105, 119–120, 162–163, 169, 170–171, 305–307, 326, 334
  for ecosystem conservation 343–344
  Farlington Marshes 320, **320**, 322
  Keyhaven marshes 194, **195**, 196
  Lower Test Nature Reserve **324**, 326
  Medmerry 333–334
  North Norfolk Coast 238, 239–240, **239**
  Outer Hebrides 103, 104
  Severn 154, 156–157, 158, 170–171
  sheep 301–302
  Solway Firth 107, 109
Great Expectations (Dickens) 179–180
Great Ouse **225**
Great Subsidence 48–49, **49**
Great Yarmouth, Norfolk 270, 279–280
Grebe, Little 226, 370
Grey, Sir Edward 287–288, **288**

Groundsel, Tree 299, 371
Groves, Henry 209–212, **209**
Groves, James 209–212, 213
Grus grus 143, 226, 268, **269**, 368
Gull, Black-headed 44, 198, **198**, 368
  Herring 262, 369
  Lesser Black-backed 262, 370
  Mediterranean 198–199, **198**, 370
Gwent Levels 151, 155–156, 160, 162–164, **162**, **163**, 167–168, 222
Gwent Wildlife Trust 164

Haematopus ostralegus 281, 370
Haines, John 'Jack' Wilton 41
Hair-grass, Somerset 170, 371
Haliaeetus albicilla 269, 271, 371
halophytes 24–26, **25**
Halstow Marshes, Kent **251**, 252
Halvergate debate 280–285
Hamford Water, Essex 310–313
Hampshire and Isle of Wight Wildlife Trust 319, 323, 327
Hare's-ear, Slender 42–43, **42**, 153, 319, 371
Harlech Castle, Gwynedd 125–126, **126**
Harrier, Marsh 313, 335, 370
Hartside Cafe, Pennine moors 56
Hatfield Chase, Humberside 230
Hawker, Southern Migrant 189, 371
hay meadows 154, **154**, 159
Hayling Island 215, **215**
Hazelwood Marshes, Suffolk 316
Heart of Darkness (Conrad) 180
Heather 82, 369
Hedera helix 121, 369
Helford River, Cornwall **286**
Helianthus annuus 163, 371
Helleborine, Dark-red 71, 369
Hemimysis anomala 301, 368
Hennediella heimii 45, **45**, 369
herbivory 22–24
  livestock see grazing, saltmeadow
  post-glacial period 140–141, 142–143
  waterfowl 22, 90–91, **91**, 109–110, 274–275, 297–298, 319–320, 322, 337
  wild ruminants 118, 119, 303–304, **304**
Hermaness, Unst 60, **60**
Heron, Grey 268, 369
Heseltine, Michael 266
Heslop Harrison, John 85, 96
Hickling Broad, Norfolk 276, 277, **277**, 278–279

Hierochloe odorata 121–123, **123**, 369
Highlands
 coastal mires **66**, 80, **81**, 82–85, **83**
 compared to English wetlands 86–87
 deer 304
 lochs 67–76, **68**, **71**, **73**, **74**, **75**
 saltmarsh 76–80, **78**, **79**
 transition zones 85–87
Hilbre Island 62, **62**
Himantopus himantopus 271, 368
Historical Flora of Middlesex (Kent) 177–178
History of British Mammals (Yalden) 118
Hollybush Common, Worcestershire 42, **42**
Holy-grass 121–123, **123**, 369
Honewort 170, 369
Hordeum marinum 224, **225**, 272, 285, **285**, 334, 370
 secalinum 281, 370
Horned-poppy, Yellow 174, **174**, 371
Horse, Wild 23, 141, 142–143, 371
Horsefly, Saltmarsh **194**, 370
Horsey Mere, Norfolk 279
Hubbard, Charles 207–208
Hubbard, John 208–209
Humber **136**; *see also* Doggerland
 Brigg prehistoric boat 147–148, 203
 coastal realignment projects 335–336
 drainage of marshes **229**, 230–231
 enclosures/embankments 223, 228–230
 Marsh Sow-thistles 148–149
 peatlands 146
 saltmarsh migration 145–148, **146**
 Stone Age peoples 137, 146–148
Huntspill Levels, Somerset 221–222
Hutcheon, Keith 73–74
Hydrobia 102
 acuta neglecta 102
Hydrocotyle ranunculoides 298, 369
 vulgaris 62, 90, 370
Hydrophilus piceus 160, 162, **162**, 369
hyper-saline conditions 16
Hythe, Hampshire **206**, **208**, 209, 210–211, 216, **216**

Ingestre Park, Staffordshire 36–38
Ingestre, Staffordshire 351

inland saltmarshes 343
 Berkshire/Oxfordshire 39–40
 Cheshire 48–53, **49**, **51**
 Gloucestershire 40–41
 inventory 347–354
 Special Areas of Conservation 289
 Staffordshire **34**, 35–38, **36**, **37**, 46–48, **47**, **254**, 256, 351
 Worcestershire 42–46, **45**
Internal Drainage Boards 280–281, 282
Inula crithmoides 127, 174, 369
invasive, native species 305–307
invasive, non-native species
 invertebrates 299–301, **300**, **301**
 mammals 302–304, **303**, **304**
 plants 297–299
Invernaver, Sutherland **19**
Iris pseudacorus 80, **80**, 332, 371
Iris, Yellow 80, **80**, 332, 371
Isle of Axholme 230–231
Itchen Ferry, Hampshire 211–212
Itchington Holt, Warwickshire 349
Iter Plantarum (Johnson) 173
Ivy 121, 369

Jackson, William 51
James I 222
Johnson, Thomas 173–177
Journal of Ecology 26–27
Juncus acutus 127, 128, **128**, 332, 371
 balticus 89–90, **90**, 368
 bufonius 275, 371
 gerardii see Rush, Saltmarsh
 maritimus 27, 334, 371

Kale, Sea 174, 371
Kalmia procumbens 71, 371
kelp 92, 369
Kenfig, Glamorgan 235
Kent, Dougie 177–178
Kentra Bay, Argyll and Bute **66**, 82–84, **83**
Ketford, Gloucestershire 41, 348
Keyhaven marshes, Hampshire **190**
 aquatic life 194, **194**
 birds 197–199, **198**
 conservation management 199
 flora 196
 grazing 194, **195**
 insects 196–197, **197**
 Lymington saltworks 193–194, **193**
Kingston Pool, Staffordshire 351
Kintail, Highlands 257, **257**
Knapdale, Argyll and Bute 78
Knapweed, Common 156, 163, 368

Knot 226, 337, 369
Koeleria vallesiana 170, 371
Kyle of Durness 86, **86**

Lactuca saligna 183–184, **183**, 291, 305, 370
Lagoon-worm, Tentacled 291, 371
lagoons 102–103, 289, 290
Laminaria spp. 92, 369
Lamprothamnium papulosum 100, 101, 102, 194, 210, 291, 369
Langstone Harbour, Hampshire 217–218, **217**, 319, 321, 322
Lapwing 169, 268–269, 271, 281, 369
Larus argentatus 262, 369
 fuscus 262, 370
 melanocephalus 198–199, **198**, 370
Lasiommata megera 196, **197**, 371
latitude 21–22, 60
Leadon River 40–41, 348
Lemming, Arctic 318, 368
 Norway 46, 370
 Siberian Brown 318, 371
Lemmus lemmus 46, 370
 sibiricus 318, 371
Lepidium latifolium 38, 272, **273**, 369
Lestes dryas 189, 370
Lettuce, Least 183–184, **183**, 291, 305, 370
Limits of the World from the East to the West, The (Asem) 192
Limnoxenus niger 160
Limonium 65, 153, 203, 272, **273**, 306, 315
 bellidifolium **28**, 370
 humile 334, 370
 vulgare 324–325, 368
Limosa spp. 226
Limosella aquatica 130, 370
 australis 129–132, 291, 371
Lindisfarne, Northumberland 235
Lion Lodge, Staffordshire 38, 351
livestock see grazing, saltmeadow
Lleidlys Cymreig 129–132, 291, 371
Local Nature Reserves (LNRs) 252–253, **253**, **254**
Loch Ainort, Isle of Skye **30**
Loch Bhornais, South Uist 97, **97**
Loch Bi, South Uist 94, 102, 103
Loch Cean a' Bhaigh, South Uist 97–98
Loch Euphort, North Uist 100
Loch Fleet, Highlands 76–77, **78**, 83, 252
Loch Gruinart, Islay **242**
Loch nam Madadh, Western Isles 101–102

Loch Nedd, Sutherland 67–73, **68**, **69, 71**, 76
Loch Nevis, Highlands 75
Loch Phaibeil, North Uist 90
Loch Sunart, Highlands 75, **75**
Loch Sween, Argyll and Bute **17**, 18, 78, **79**
London 20, **21**, 177–178, 180; see also Thames estuary
Lotus spp. 156–157
Lousewort, Marsh 90, 370
Lower Test Nature Reserve, Hampshire 323–327, **323, 324, 325, 326**
Luachair Bhailtigeach 89–90, **90**, 368
Lucanus cervus 162, 371
Lus na Peighinn 62, 90, 370
Lus na Saillteachd see Sea-milkwort
Lus nan Leac 71, 84, 210, 369
Lus Riabhach 90, 370
Lutra lutra 142, 370
Lymington River 132, 198–199, 212, 213, **213**, 327
Lymington saltworks, Hampshire 193–194, **193**
Lynx 142, 370
Lynx lynx 142, 370
Lysichiton americanus **296**, 299, 368

machair **88**, 89, **93, 94, 95**
  farming 103–105, **104, 105**
  overview 92, 94–95
  pondweeds 96–99, **97**
Macoma balthica 337, 368
Mallard 143, 370
Malvern Hills 42–46, **42, 45**
Mammoth 46, 141, 276, 370
Mammuthus primigenius 46, 141, 276, 370
Marcham Marsh, Berkshire/ Oxfordshire 39–40, 347–348
Marsh-mallow 174, 223–224, **224**, 272, 285, 370
Marsh-orchid 52, 90, 332, 370
  Early 327, 369
Mawddach estuary **124**
McLachan, Margaret 123, **123**
Meadowsweet 80, 370
Medmerry, Sussex 294–295, **294**, 333–334, **333, 334**
Mehalah, a story of the salt marshes (Baring-Gould) 181
Mercenaria mercenaria 299–300, 369
Mergus albellus 269, 371
  merganser 226, 369
Mersehead, Dumfries **108**, 120
Metamorphosis of Plants, The (Goethe) 24

Metrioptera roeselii 196, 197, 370
Michaelmas-daisy, American 299
Mickletown Ings, Yorkshire 354
Middlewich, Cheshire 48, 354
Midge, Meniscus 162, 370
Migratory Birds Treaty 287
military sites 259, **259, 293**
Millennium Ecosystem Assessment, U.N. 293–295
Ministry of Agriculture 280, 282–283
Mistletoe 133, 370
mobility of marshes 323, **325**
Moenchia erecta 196, 371
Mogairlean 52, 90, 332, 370
Mòine Mòhr, Argyll and Bute 80, **81**, 82, **82**, 86–87, **87**
Montagu of Beaulieu, John Douglas-Scott-Montagu, 2nd Baron 201–202
Morecambe Bay 128
Morfa Harlech, Gwynedd 125–127, **126, 127**, 130, 252
Morvich Marsh, Highlands 73–75, **74**, 257
Mossman, Hannah 332–333
Motacilla flava 281, 371
Moth, Fisher's Estuarine 291, 309–313, **310, 311**, 369
Mound Alderwoods, Highlands 76–78, **77**
mud-snails 102
Mudwort 130, 370
  Welsh 129–132, 291, 371
Mussel
  Quagga 300, 370
  Zebra 300, 371
Myriophyllum 95, **95**, 371
  spicatum 62, 281, 371
Mysid, Bloody-red 301, 368

Naiad, Holly-leaved 276–278, **278**, 291, 369
Najas marina 276, 277–278, **278**, 291, 369
Nantwich, Cheshire 48, 49, 51, 352–353
Napton, Warwickshire 350
National Nature Reserves (NNRs) 50, 164, **175**, 244, 245–246, 247, 251–252
National Parks and Access to the Countryside Act (1949) 247
National Trust 233–234, 236–237, 240, 255, 256–257, 257–258, **257**, 329, 330, 332
National Vegetation Classification of British Plant Communities 31–33

Natural History (Pliny the Elder) 133–134
Nature Conservancy Council 187, 251, 265, 266, 267, 268, 282, 283
Nature Conservation Review, A (Ratcliffe) 247–250
neap tides 17
Needles Headland, Isle of Wight 63, **63**
Nematostella vectensis 194, **194**, 291, 371
Neòinean Cladaich see Thrift
neophytes 203
Nettle, Common 240, 368
Neumann's Flash, Cheshire 51–52, **51**, 353
New Forest 114, 191, 196, 199, 299
Newent, Gloucestershire 348
Newtown, Isle of Wight 203, **204**, 215, 257–258
Norfolk Naturalists' Trust 236, 238
Norfolk Wildlife Trust 238
North Norfolk Coast **55**; see also Blakeney Point, Norfolk
  carbon sequestration 294
  conservation 236–237, 238
  saltmeadow grazing 239–240
  storm of 1953 313
  Tansley's descriptions 26–29
  Titchwell Nature Reserve **308**, 313–315, **314, 316**
North Ronaldsay, Orkney **303**
North Uist/Uibhist a Tuath 90–91, 100, 245
North Wales Wildlife Trust 131–132
Northey Island, Essex 329
Northwich, Cheshire 48, 49, 52, 353
Nowers, John 46–47
Numenius arquata 169, 337, **337**, 369

Oenanthe crocata 80, 369
  lachenalii 39, 43, **128**, 148, 153, 171, 196, 271–272, 370
Oliver, Francis 26–27, 233–234, 236, 238
Orache 56, 186, 298, 370
  Babington's 368
  Early 86, 369
  Grass-leaved 57, 369
  Long-stalked 86, 370
  Shrubby 299, 371
  Spear-leaved 45, 86, 371
Orange moth, Frosted 369
Orchid, Green-winged 326, 369

Orconectes limosus 300, 371
Orton Pits, Berkshire 348
Orton Pits, Cambridgeshire 101
Osmunda regalis 196, 370
Osprey 226, 271, 370
Ostrea edulis 299, 368
Otter 142, 370
Our Mutual Friend (Dickens) 180
Ouse River (Yorkshire) 146, 148–
149, **149**, 230; see also Humber
Outer Hebrides 89
Blacklands 94, **94**
driftwood 98–99
machair **88**, 92, **93**, 94–95, **94**, **95**
Newton Estate 251
pondweeds 96–99, **97**
Red Deer 304
salinity 90, 94
stoneworts 99–101, 101–102
Ovis aries 128, 238–239, 302–303,
303, 369
orientalis 302, 371
Owl, Short-eared 97, 334, **335**, 371
Oyster, Common 299, 368
Oystercatcher 281, 370

Paget, Charles 270–272
Paget, Sir James 270–272, **271**
Pallis, Marietta 278–279, **279**
Pandion haliaetus 226, 271, 370
Panurus biarmicus 313, **314**, 335, 368
Paracymus aeneus 291, 368
Pasturefields, Staffordshire **34**,
35–38, **36**, **37**, 256, 351
Pauntley, Gloucestershire 40–41
Payford Bridge, Gloucestershire 41
peatlands 82, **82**, 86–87, 135, 145,
146, 166–167, 270
Pedicularis palustris 90, 370
Pelecanus crispus 223, 369
Pelican, Dalmatian 223, 369
Pennine moors 56
Pennywort, Floating 298, 369
Marsh 62, 90, 370
perched saltmarshes 61–63, **61**, **62**,
63, 244, 252, 289
Persicaria minor 326, 371
mitis 326, 371
Peucedanum officinale 175–176, **176**,
311–312, **311**, 369
Philomachus pugnax 269, 370
Phoenicopterus chilensis 191, 368
Phragmites australis 24, 163, 240,
279, 297, 305, 333, 335–336,
368
Phytophthora alni fungus 78
Pigmyweed, New Zealand 298, 370
Pintail, Northern 269, 370
Pipewort 84, 370

Plantago maritima 27, 35, 46, 56, 61,
69, 71, 73, 140, 371
Plantain, Sea 27, 35, 46, 56, 61, 69,
71, 73, 140, 371
Platalea leucorodia 269, **328**, 340,
371
Plot, Dr Robert 36–37
Plover, Golden 169, 271, 369
Pluvialis apricaria 169, 271, 369
Poly-Olbion (Drayton) 225–226
Polypogon monspeliensis 319, **319**,
368
Pond-sedge, Greater 326, 369
Pondweed 96–99, **97**, 102, 277
American 97–98, 368
Fennel 96, 101, 278, 281, 369
Grass-wrack 369
Horned 52, 369
Lesser 96, 101, 370
Shetland 96–97, 371
Slender-leaved 97, 371
Pool Hill, Gloucestershire 41
Poole Harbour, Dorset 207, 213–
214, 303–304, **304**
pools 31, **32**
Porlock Weir, Somerset 235
Porter, Endymion 228–229, **228**
Portsmouth, Hampshire 179, 321,
321
post-glacial period, immediate
22–23, 142
Potamogeton 96–99, **97**, 102, 277
compressus 369
epihydrus 97–98, 368
filiformis 97, 371
pectinatus 96, 101, 278, 281, 369
pusillus 96, 101, 370
rutilus 96–97, 371
Potentilla anserina 62, **88**, 90–92,
91, 371
Pottia, Heim's 45, **45**, 369
prehistoric flora/fauna 22–23,
45–46, 116–117, 118–119,
135, 276, 302, 304; see also
Doggerland
prehistoric human populations 119,
137–138, 143–144, 146–148,
302, 304
Primrose, Scottish 61, **61**, 370
Primula scotica 61, **61**, 370
veris 327, 369
Procambarus clarkia 300, **301**, 370
protected species 290, 291
Protection of Birds Act (1954) 319
Puccinellia 303–304, 319, 320
distans 44–45, 49, 52, 53, 55, 325,
370
fasciculata 38, 368
maritima 28, 49, 68, 126, 153,

239, 305, 315, 337, 368
rupestris 52, 325, 371
Puffin **60**, 226, 370
Pulicaria dysenterica 163, 369
Puxton, Somerset 151, **152**, 155,
159

Quaking-grass 326–327, 370
quality 343
Queen of Love, The (Baring-Gould)
48, 50

rainfall 21, 90, 127, 129, 184
Ramsar Convention 288, 292
Ramsar sites 292, 319
Rangifer tarandus 46, **140**, 141, 370
Ranunculus baudotii 47, **47**, 368
ophioglossifolius 211, 368
sceleratus 45, **45**, 298, 368
Ratcliffe, Derek 61, 247
Raven, John 85
Raven's Point, Ireland 235
Ray Island, Essex 181, 240
Ray, John 130, 223–224, **223**
readjustment, isostatic 19
Recurvirostra avosetta **34**, 43–44,
269, 313, 315, 368
Red Data Book 130, 305
Redbridge, Hampshire **323**
Redmarley D'Abitot,
Gloucestershire 40–41
Redshank 265–266, 281, 370
Reed, Common 24, 163, 240, 279,
297, 305, 333, 335–336, 368
Reed-mace, Lesser 279, 370
regional differences 28
Reindeer 46, **140**, 141, 370
rejuvenation 329
relative sea levels 18–21, **21**
Rhinoceros, Woolly 46, 276, 371
Rhynchospora fusca 84, 368
Rio de Janiero Convention on
Biological Diversity 288
road verges 53–57, **54**, **57**
Rodel, Isle of Harris **12**
Rodwell, John 31–32
Roman period 43, 148–149, 155–
156, 180, 279–280, 303
Romney Marshes 235, 240–241
Roosevelt, Theodore 287–288, **288**
Rostherne Mere, Cheshire 50
Rostonstown Burrow, Ireland 235
Rothschild, Charles 233–235, **234**,
240
Royal Society for the Protection of
Birds (RSPB) 255–256
Arne **256**
Avocet emblem 44
Gedney Drove End 265, 266

Loch Gruinart **242**
Meddatt Marsh 336–337, **337**
Medmerry 333
Mersehead 120
Newport Wetlands 164
Romney Marsh 240–241
Titchwell 313–315
Wallasea 338–341, **339**, **340**
Rudder, Samuel 40–41
Ruff 269, 370
Rùm 85
Rumex maritimus 47, 48, 369
    rupestris 64–65, **65**, 290, 291,
    305, 371
Ruppia spp. 102, 281, 317, **317**, 371
Rush, Baltic 89–90, **90**, 368
    Saltmarsh 278, **285**, 370
        the Broads 284–285
        cliffs 60
        inland saltmarshes 35, 38, 39,
        40, 41, 47, **47**, 48
        Lower Test Nature Reserve
        324, 326
        Outer Hebrides 90
        Pennine moors 56
        perched saltmarshes 62
        Roman York 148
        Severn 153
        Solent 207
    Sea 27, 334, 371
    Sharp 127, 128, **128**, 332, 371
    Toad 275, 371

Saiga tatarica 141, 370
Saldon near Himbledon,
    Worcestershire 349
Salicornia; see also glassworts
    ramosissima 52, 370
salinity 16–17, 44, 90, 94, 157–158,
    160, 276, 278–280
Salisbury, Robert 59
Salix 325, **325**
    herbacea 46, 369
salt industry 37–38, 43, 48–50, **49**,
    52, **53**, 193–194, **193**
salt spreading on roads 53–57, **54**,
    57
salt, wind-borne
    cliff pools 63–64, **64**
    cliff tops 60–61, **60**
    Outer Hebrides 90
    perched saltmarshes 61–63, **61**,
    62, **63**
    Salisbury's account 59
    storm beach saltmarshes 64–65,
    65
Salthouse, Norfolk 236
Saltmarsh-grass 303–304, 319, 320
    Borrer's 38, 368

Common 28, 49, 68, 126, 153,
    239, 305, 315, 337, 368
Reflexed 44–45, 49, 52, 53, 55,
    325, 370
Stiff 52, 325, 371
Samolus valerandi see Brookweed
samphire 186, 238, 279, 330
Sandbach Flashes, Cheshire 354
Sandhurst, Gloucestershire 40
Sandworm, Lagoon 291, 369
Sarcocornia spp. see glassworts
Saul Warth, Gloucestershire 169
Scheonus nigricans 70, **70**, 368
Schoenoplectus tabernaemontani 38,
    41, 45, 48, 53, 369
    triqueter 178–179, **178**, 291, 305,
    371
Scolt Head Island, Norfolk 29, 236,
    238
Scott, Nick 56
Scott, Sir Peter 255
Scottish Saltmarsh Survey (Haynes)
    60–61
Scottish Wildlife Trust 120
Scurvygrass 54, 69–70
    Atlantic 69–70, 368
    Common 54–55, **54**, 61, 69–70,
    368
    Danish 54, 57, **57**, 369
    English 153, 169, 177–178, 271,
    369
    Scottish 69–70, 370
Sea-anemone, Ivell's 290, 291, 369
    Starlet 194, **194**, 291, 371
Sea-blite, Annual 63, 174, 207, 281,
    332, 368
    Shrubby 27, 55–56, **55**, 371
sea defences; see also coastal
    engineering; embankments
    costs 294–295
    failed 294–295, 329–330, **331**,
    332
    realignment projects see coastal
    realignment projects
    Severn 151–152, 156, 157, 164,
    167–171, **168**
    Thames estuary 182–183
Sea-heath 27, 63, **63**, 224, **290**,
    334, 371
Sea-lavender 65, 153, 203, 272,
    273, 306, 315
    Common 324–325, 368
    Lax-flowered 334, 370
    Matted **28**, 370
sea loch marshes 29, **30**
Sea-mat, Trembling 291, 371
Sea-milkwort 371
    Bae Ceredigion 126
    Brigg boat find, Humber 147

the Broads 271
Highland marshes 69, 73
inland saltmarshes 36, 37–38, 46,
    47, 53
North Norfolk Coast **28**
Pennine moors 56
perched saltmarshes 62
road verges 55
Sea-purslane 24, 371
    the Broads 272
    coastal realignment projects
    333, 334
    declining saltmarshes 24, 305,
    339
    Lower Test Nature Reserve
    324
    North Norfolk Coast 27,
    238–239
    perched saltmarshes 63
    Severn 153
    the Solent 199, 207
    Thames estuary 174
    Pedunculate 186–188, **187**, 271,
    291, 305, 370
Sea Slug, Lagoon 291, 369
Sea-spurrey, Lesser 36, 43, 44, 49,
    53, 55, 223–224, 275, 325, 370
Sedge, Curved 55, 61, 72–73, **72**,
    73, 305, 369
    Distant 38, 39, 40, 41, 56, 157,
    369
    Divided 157, 271, 326, **326**, 369
    Dotted 196, 369
    Estuarine 76, 369
    Hair 86, 369
    Long-bracted 65, 370
    Saltmarsh 73–76, **74**, 257, 370
Sèimhean Dubh 70, **70**, 368
Seisg Bheag Dhubhcheannach 55,
    61, 72–73, **72**, **73**, 305, 369
Seisg nam Measan Beaga 61, 69, **69**,
    73, 371
Seisg Rèisg Ghoirt 46, 61, 69, **69**,
    82, 370
Severn
    borrow dykes **161**
    consequences of sea defences
    165–167
    Dolemoors 159–160
    drainage of marshes 158, **158**
    embankments 156, 165, 166, 167,
    170
    enclosures 154, 160, 221–222
    floods **150**, 151, 167
    hay meadows 154, **154**, 157, 159,
    164
    insects 160, 162–164, **162**, **163**
    livestock grazing 154, 156–157,
    158, 159, **161**, 162–163, 164,

169, 170–171, 222
marshes behind sea defences
   156–160
peatlands 167
salinity 157–158, 160
sea defences 151–152, 155–156,
   157, 159, 164, 167–171, **168**
tides 18
trenches **31**
Warths and Dumbles 153
Wildfowl and Wetlands Trust 255
Share Marshes, Suffolk **284**
Sheep, Domestic 128, 238–239,
   302–303, **303**, 369
   Urial 302, 371
Shelduck 143, 371
Sheppey, Kent **172**, 173
Shin River 235
Shipworm 98, 371
Shirleywich, Staffordshire 37–38,
   351
Shoard, Marion 267
Shrimp, Brine 191–193, **192**, 199,
   368
   Killer 301, 369
   Lagoon Sand 194, 291, 369
   Tadpole 112–116, **113**, 123, 291,
      371
Silene viscaria 121, **121**, 371
Silver Spring, Cheshire 354
Silverweed 62, **88**, 90–92, **91**, 371
Sites of Special Scientific Interest
   (SSSIs) 241, 244, 247, 265,
   267–268, 282, 288–290, 319,
   320–321
Skunk-cabbage, American **296**,
   299, 368
Slàn-lus na Mara see Plantain, Sea
Slimbridge, Gloucestershire 165–
   166, **165**, 255
sluices 18, 76–77, **77**, 167
Smew 269, 371
Smith, Mike 47–48
Snail, De Folin's Lagoon 291, 369
Snàthainn Bhàthaidh 95, **95**, 371
Snipe, Common 271, 281, 369
Society for the Promotion of
   Nature Reserves 234–236, 237,
   240, 243
Society of Apothecaries 177
Solent, the 235, 253, 289, 294,
   297–298, 299, 319–322, 327;
   see also Keyhaven marshes,
   Hampshire; Southampton
   Water *Spartinas*
Solway Firth **109**
   agriculture 107, 109
   Aurochs and Elk 118–120
   Barnacle Geese 109–112, **110**

Edward I 107, **107**
environmental stressors 112
merse **106**
Natterjack Toads 116–117, 120
political history 107
Southwick 120–123, **120**, **122**
Tadpole Shrimps 112–116
Wigtown martyrs 123, **123**
Somerset Levels 151, 152, 157–
   158, **166**, 167, 221–222
Sonchus palustris 148–149, **149**, 178,
   224, 272, 370
South Hams, Devon **20**, **65**, 330
South Uist/Uibhist a Deas 90, 94,
   97, **105**, 244, 245, 249
Southam Holt, Warwickshire 350
Southampton Water *Spartinas* **215**,
   216
   current status 214–219, **215**,
      216, **217**, **218**
   expansion/contraction 212–214,
      213
   further hybridisation 217
   global spread 202–203
   Groves brothers' work 209–212
   hybridisation 201, 204–205, **205**,
      207–209
   Montagu's promotion 201–202
   S. alterniflora 201, 204–207, **205**,
      206
   S. anglica **200**, 201, **202**, **208**
   S. maritima 201, 203–204, **204**,
      205
   S. × townsendi 201, 211–212
Southwick, Dumfries and Galloway
   115, 120–123, **120**, **122**
Sow-thistle, Marsh 148–149, **149**,
   178, 224, 272, 370
Spartina 303–304
   alterniflora 201, 204–207, **205**,
      206, 211–212, 215–216, **216**,
      371
   anglica see Cord-grass, Common
   maritima see Cord-grass, Small
   patens 218, **218**, 299, 370
   × *townsendii* 28, 201, 207–208,
      210–211, 212, **216**, 217, 371
   versicolor 219
Spearwort, Adder's-tongue 211,
   368
Special Areas of Conservation 289,
   355–360
Special Protection Areas 292, 319
Spergularia marina 36, 43, 44, 49,
   53, 55, 223–224, 275, 325, 370
Spike-rush, Common 148, 369
   Dwarf 132, **133**, 291, 369
   Slender 140, 371
Spoonbill 269, **328**, 340, 371

spring tides 17
Spurge, Sea 174, 371
Staffordshire Wildlife Trust 35
Stanpit Marsh LNR, Dorset **254**
Starling 336, 371
Steart peninsula, Somerset 169–
   170, **170**, 171
Stilt, Black-winged 271, 368
Stone Age 118, 138, 143–144, 302,
   304; see also Doggerland
Stonewort 53, 99–102, 277, 278
   Baltic 101–102, 277, 368
   Bearded 101, **101**, 277, 291, 368
   Foxtail 100, 101, 102, 194, 210,
      291, 369
   Rough 98, 100, 101, 102, 277,
      370
Stork, Black 143, 368
   White 143, 269, 271, 371
storm beach saltmarshes 64–65, **65**
storm-surge events 18, **18**
strandlines **23**, 24, **24**
Stratiomys longicornis 64, **64**, 370
   singularior 162, **162**, 369
Sturnus vulgaris 336, 371
Suaeda maritima 63, 174, 207, 281,
   332, 368
   vera 27, 55–56, **55**, 371
subsidence 48–51, **49**
Suffolk Wildlife Trust 284–285,
   284, 316
Summer-cypress 56–57, **57**, 299,
   371
Sundew, Oblong-leaved 84, 370
Sunflower 163, 371
Sus scro 142, 371
Swan, Bewick's 269, 281, 368
   Whooper 143, 371
Sweet-grass, Reed 326, 370

Tachybaptus ruficollis 226, 370
Tadorna tadorna 143, 371
Tain Sandhills, Ross 235, **236**
Tamarisk 299, 371
Tamarix gallica 299, 371
Tansley, Sir Arthur 26–28, **27**, 31,
   238
Taraxacum spp. 71, 369
tasselweeds 102, 281, 317, **317**,
   371
Taxus baccata 147, 371
Taylor, George 96
Taynish, Argyll and Bute **17**, 252
Teal, Eurasian 226, **226**, 269, 369
Teasel, Wild 164, 371
Tenellia adspersa 291, 369
Tephroseris palustris ssp *palustris*
   272, 370
Teredo navalis 98, 371

Thames estuary
  chenopods 184–186
  conservation 235
  dragonflies and damselflies 188–189
  Johnson's records 173–177
  livestock grazing 182–183
  other apothecaries' records 177–179
  Pedunculate Sea-purslane 186–188
  popular images 179–181
  sea defences 182, 183
  Storm of 1953 188, **189**
  Wicks 182–183, **182**
Theft of the Countryside, The (Shoard) 267
Thetidia smaragdaria 240, 369
Thrift 371
  Broads absentee 272, **274**
  cliffs 61
  coastal realignment projects 334
  Highlands 69, 71, 73, 80, 83, **83**
  Keyhaven marshes 196
  perched saltmarshes 63
  Severn 153
Thurne River 274, 276, 277, 278, 279
tidal ranges 17–18, **18**
Tit, Bearded 313, **314**, 335, 368
Titchwell RSPB Nature Reserve, Norfolk 256, **308**, 313–315, **314**, **316**
Tixall, Staffordshire 37, 38, 351
Toad, Natterjack 115, 116–117, **117**, 118, 120, 291, 370
Torrisdale Bay, Highlands 75
Townsend, Frederick 205–206, 210
Traeth Mawr, Gwynedd 130, 131–132, **131**
trefoils, bird's-foot 156–157
trenches **31**
Trent 137–138, 146, 230, 335–336; *see also* Branston, Staffordshire; Humber; Pasturefields, Staffordshire
Trifolium 110, 157, 163, 196
  fragiferum 40, 148, 153, 171, 174, 371
  ornithopodioides 196, 368
  squamosum 64, 157, **159**, 171, 178, 272, 319, 370
  striatum 196, 369
  subterraneum 196, 371
Triglochin 320
  maritimum see Arrowgrass, Sea

Tringa totanus 265–266, 281, 370
Trinia glauca 170, 369
Triops cancriformis 112–116, **113**, 123, 291, 371
Typha angustifolia 279, 370

Uibhist a Deas/South Uist 90, 94, 97, **105**, 244, 245, 249
Uibhist a Tuath/North Uist 90–91, 100, 245
UNESCO sites 288
United Kingdom National Ecosystem Assessment 294–295
Upton Warren flashes, Worcestershire 43–46, **45**, 349
urban development 322
Ursus arctos 142, 368
  maritimus 112, 370
Urtica dioica 240, 368
Usan, Angus 61, **61**

Vanellus vanellus 169, 268–269, 271, 281, 369
vegetation, distribution of 21–22
vegetation traps **29**
Vermuyden, Cornelius 230
Vernal-grass, Sweet 122, 156, 371
Victorella pavida 291, 371
Viscum album 133, 370
Voyage of the Beagle, The (Darwin) 191–192

Wagtail, Yellow 281, 371
Walborough Marshes, Somerset 170–171, **171**
Wall butterfly 196, **197**, 371
Wallasea, Essex 338–341, **339**, **340**
Walter of Henley 13, 15
Warbler, Cetti's 325, 368
  Reed 335–336, 370
Ware Park Brickfields, Hertfordshire 347
Warming, Eugenius 24–25
warping 230
Wash, the
  17th-century 222–224, 225–226
  19th-century 224
  birdlife **220**, 225–226, **227**, **228**
  common rights 222–223
  development proposals 261–262, 263
  embankments 262–263
  Gedney Drove End **260**, 263–268, **264**
Spartinas 202

SSSIs 265, 289
Water-crowfoot, Brackish 47, **47**, 368
Water-dropwort, Hemlock 80, 369
  Parsley 39, 43, **128**, 148, 153, 171, 196, 271–272, 370
Water-milfoil 95, **95**, 371
  Spiked 62, 281, 371
Water-pepper
  Small 326, 371
  Tasteless 326, 371
waterfowl grazing 22, 90–91, **91**, 109–110, 274–275, 297–298, 319–320, 322, 337
Waterwort, Eight-stamened 147–148, 369
Watson, Hewett Cottrell 211
Wells, James 46–47
West Norfolk Plants Today (Petch and Swann) 239–240
White Sea, Russia 140
Whitlowgrass, Hoary 46, 369
Wick River 235
Widewater Lagoon, Sussex 290, **290**
Wigeon, Eurasian 143, 169, 226, 281, 284, 369
Wigtown Bay LNR 253, **253**
Wildfowl and Wetlands Trust 165, 170, 171, 255
Wildlife and Countryside Act (1981) 267–268, 282, 288, 291
Wildlife Trusts 243, 255, 256
Willow 325, **325**
  Dwarf 46, 369
Wilson, Margaret 123, **123**
Winsford, Cheshire 52, **53**, 352
Witton Flash, Cheshire 48–49, **49**
Wolf 142, 371
woodlands 76–78, **78**, **79**, 80, 120, 134–135
Worcestershire Wildlife Trust 43
Wormwood, Sea 127, **128**, 171, 174, 240, 271, 306, 371
Wrack, Horned 102, 369

Yalden, Derek 22, 118, 141, 302
Yellow-sedge, Small-fruited 61, 69, **69**, 73, 371
Yew 147, 371
Ynyslas, Ceredigion 28
York 148–149, **149**

Zannichellia palustris 52, 369
Zostera spp. 27, 68, 92, 102, 214, 217, 319, 369